CAMBRIDGE LIBRARY COLLECTION

Books of enduring scholarly value

Astronomy

From ancient times, humans have tried to understand the workings of the world
around them. The roots of modern physical science go back to the very earliest
mechanical devices such as levers and rollers, the mixing of paints and dyes,
and the importance of the heavenly bodies in early religious observance and
navigation. The physical sciences as we know them today began to emerge as
independent academic subjects during the early modern period, in the work of
Newton and other 'natural philosophers', and numerous sub-disciplines developed
during the centuries that followed. This part of the Cambridge Library Collection
is devoted to landmark publications in this area which will be of interest to
historians of science concerned with individual scientists, particular discoveries,
and advances in scientific method, or with the establishment and development
of scientific institutions around the world.

A Popular Guide to the Heavens

A talented mathematician trained at Trinity College, Dublin, Sir Robert Stawell Ball
(1840–1913) was best known in the early twentieth century for his immensely
popular books on astronomy. He also gave the Royal Institution's Christmas Lectures
on five occasions. First published in 1905, this concise guide to the basics of
astronomy assumes almost no prior knowledge of the subject. Beginning with
simple phenomena such as the seasons and the effects of atmospheric refraction,
Ball expands quickly into month-by-month indexes of the night sky, star charts,
and explanations of some of the lesser-known stellar and solar features, from the
paths of sunspots to details of the major nebulae. Including over eighty pages
of meticulous charts and illustrations, his book remains an excellent resource for
students in the history of science, and interested laypeople. Also reissued in this
series are *The Story of the Heavens* (1885) and *Star-Land* (1889), alongside Ball's
more technical *Treatise on Spherical Astronomy* (1908).

Cambridge University Press has long been a pioneer in the reissuing of out-of-print titles from its own backlist, producing digital reprints of books that are still sought after by scholars and students but could not be reprinted economically using traditional technology. The Cambridge Library Collection extends this activity to a wider range of books which are still of importance to researchers and professionals, either for the source material they contain, or as landmarks in the history of their academic discipline.

Drawing from the world-renowned collections in the Cambridge University Library and other partner libraries, and guided by the advice of experts in each subject area, Cambridge University Press is using state-of-the-art scanning machines in its own Printing House to capture the content of each book selected for inclusion. The files are processed to give a consistently clear, crisp image, and the books finished to the high quality standard for which the Press is recognised around the world. The latest print-on-demand technology ensures that the books will remain available indefinitely, and that orders for single or multiple copies can quickly be supplied.

The Cambridge Library Collection brings back to life books of enduring scholarly value (including out-of-copyright works originally issued by other publishers) across a wide range of disciplines in the humanities and social sciences and in science and technology.

A Popular Guide
to the Heavens

ROBERT STAWELL BALL

CAMBRIDGE
UNIVERSITY PRESS

University Printing House, Cambridge, CB2 8BS, United Kingdom

Published in the United States of America by Cambridge University Press, New York

Cambridge University Press is part of the University of Cambridge.
It furthers the University's mission by disseminating knowledge in the pursuit of
education, learning and research at the highest international levels of excellence.

www.cambridge.org
Information on this title: www.cambridge.org/9781108066495

© in this compilation Cambridge University Press 2013

This edition first published 1905
This digitally printed version 2013

ISBN 978-1-108-06649-5 Paperback

A POPULAR GUIDE

TO THE

HEAVENS.

A

POPULAR GUIDE

TO

THE HEAVENS

A SERIES OF EIGHTY-THREE PLATES,

WITH EXPLANATORY TEXT & INDEX.

BY

SIR ROBERT STAWELL BALL, LL.D., F.R.S.,

LOWNDEAN PROFESSOR OF ASTRONOMY AND GEOMETRY IN THE UNIVERSITY OF CAMBRIDGE.

GEORGE PHILIP & SON, Ltd.

LONDON: THE LONDON GEOGRAPHICAL INSTITUTE, 32 FLEET STREET, E.C.

LIVERPOOL: PHILIP, SON & NEPHEW, 45-51 SOUTH CASTLE STREET.

1905.

PREFACE.

THE object of the present work is to provide a popular guide to the study of the sky by furnishing a summary of our present knowledge of the Solar system, a guide to the positions of the planets for the first half of the present century, a series of star maps, some examples of the finest achievements in the art of drawing and photographing celestial objects, and a list of interesting objects which may be observed with small telescopes.

In the text will be found a descriptive account of the plates, and of the methods of using the maps and tables. It is, however, desirable to draw attention in this place to certain characteristics of the work, and to make my acknowledgment to the friends who have kindly assisted me.

In 1892 I edited an atlas of the celestial bodies which has long been out of print. Though this atlas bore my name yet as explained in the preface it was largely due to my friend Dr. Rambaut. The question of a new issue of the work having arisen, it was deemed better to recast the book completely, and the present volume is the result.

The star maps carefully drawn by Dr. Rambaut having been corrected for the changes obviously required by the lapse of 12 years have been retained, so have also the maps of the Moon, drawn by the late Mr. Elger.

But the advance of Astronomical portraiture has rendered it necessary to supersede most of the remaining plates by new material. This has involved so many changes in the text that the book is substantially a new one, and is now arranged in such a manner as may, I hope, entitle the book to be called a Popular Guide to the Heavens.

The map of Mars is reduced from the large map made by Mr. A. E. Douglass, published in Volume II. of the Annals of the Lowell Observatory.

The drawings of Jupiter on plate 9 have been copied from Dr. O. Lohse's observations, in the third volume of the *Astrophysiklische Observatorium* at Potsdam. The drawings of Jupiter's satellite are from a paper by Professor Barnard, published in the *Monthly Notices of the Royal Astronomical Society*. The drawing of Saturn, plate 10, by Professor Barnard, is reproduced from a drawing also published in the *Monthly Notices*.

The photograph of the great sunspot of 1898 September is reproduced from a photograph taken at the Royal Observatory, Greenwich, for which I am indebted to the kindness of the Astronomer Royal.

The drawings of the Solar Prominences on plate 12 are from a paper by Herr Fenyi, in the *Astrophysical Journal*; the picture of the Solar Prominence photographed by Professor Barnard is from the report of the Yerkes Observatory Eclipse Expedition to Wadesboro', U.S.A., published in the same journal.

The map of the paths of Solar Eclipses, 1901-1950, has been prepared from the series of maps in Dr. Oppolzer's great work, *Canon der Finsternisse*.

The photographs of typical Solar Coronæ on plate 16 are selected from the series of Eclipse photographs brought together by the Royal Astronomical Society.

Plate 17, the drawing of Donati's Comet, by the late Professor Bond, is from the splendid volume of observations of that comet in the *Annals of the Harvard College Observatory*.

I am indebted to Professor Barnard for the use of the comet photographs in plate 18, which are selected from the series of his photographs published by the Royal Astronomical Society. I owe to him also the photograph of a region of the Milky Way in plate 81, taken from the same series.

For the permission to reproduce the two comet photographs of plate 19, I must thank the Astronomer Royal of England and the Astronomer Royal at the Cape, respectively.

I am greatly indebted to Professor Hale, Director of the Yerkes Observatory of the University of Chicago, and to Mr. Ritchey, Astronomer at that observatory, who took the photographs, for permission to use the three photographs of the Moon, plates 20, 21, 22; the photographs of the nebulæ in Orion and Andromeda, plates 73 and 47, and the drawings of the nebulæ round Nova Persei. The photographs of that star and of the region in which it appeared I owe to Mr. Stanley Williams, of Brighton.

My friend Mr. W. E. Wilson kindly allows me to use the photographs of the Cluster in Hercules, and of the nebula in Cygnus, forming plate 75, which were taken by him at his observatory at Daramona, county Westmeath. To Professor Campbell, Director of the Lick Observatory, I owe permission to reproduce the three photographs of the spiral nebula in Canes Venatici, the Ring nebula in Lyra, and the Dumbbell nebula in Vulpecula, made with the Crossley Reflector by his lamented predecessor, Professor J. E. Keeler.

Plate 80, of the Pleiades Cluster, is taken from a photograph by the brothers Henry, published in a report of the Paris Observatory; and for the photograph of the nebulæ in that cluster I am indebted to the late Dr. Isaac Roberts, of Crowborough, Sussex.

Plate 83, illustrating the adoption of Standard Time, and the line where the date changes, has been made from information kindly furnished by the Hydrographer to the Admiralty.

The work involved has been very onerous, and I could not have undertaken it had I not been so fortunate as to have had the aid of Mr. Arthur Hinks, M.A., Chief Assistant at the Cambridge Observatory. To Mr. Hinks I am indebted for the selection of the new plates, as well as for the preparation of the text which accompanies them. I would like to record my thanks to him for all his skill and zeal.

CAMBRIDGE,   ROBERT S. BALL.
November, 1904.

CONTENTS.

INDEX.

LIST OF ILLUSTRATIONS.

CHAPTER I.

PLATES 1 & 2.

THE CELESTIAL SPHERE, THE SEASONS, &c.

Our first plate is an attempt to represent in a diagram what cannot properly be figured, save in the imagination of the student, the relation of the terrestrial sphere to the celestial. The celestial sphere must be *imagined* of infinite diameter; in our figure it has to be *represented* as only twice the diameter of the orbit of the Earth about the Sun.

The centre of the Earth describes an ellipse, very nearly a circle, in a plane which passes through the centre of the Sun. If this plane is produced all ways to infinity it cuts the celestial sphere in a great circle—the Ecliptic. Could the Sun be viewed among the stars from the centre of the Earth, it would be seen to lie always upon this great circle.

The axis of rotation of the Earth makes a constant angle with the plane of the ecliptic, and points constantly in the same direction. (We are neglecting here the very slow effects of precession). Consequently the plane of the Earth's equator, produced all ways to infinity, cuts the celestial sphere in a fixed great circle which is the celestial equator, and the axis of rotation of the Earth, similarly produced to infinity, cuts the celestial sphere in two points which are the poles of the celestial equator, more commonly called the poles of the sky.

It is now easy to see that at northern midwinter the north pole of the Earth is turned away from the Sun, which is at the winter solstice, the point where the ecliptic is farthest south of the celestial equator. As spring advances, the Sun, apparently moving along the ecliptic, approaches the celestial equator, and enters it at the vernal equinox, otherwise known as the "first point of Aries." At northern midsummer the north pole of the Earth is turned towards the Sun, which then appears farthest north of the celestial equator, and thence forward begins to dip again towards the autumnal equinox.

It is found convenient to refer the places of all stars on the celestial sphere to the celestial pole and equator. Distance north or south of the equator is called north or south declination, and corresponds to north or south latitude of a place on Earth. This provides for one co-ordinate. The other, called Right Ascension in the sky, corresponds to longitude on Earth; and as the longitude of a place is measured from a meridian on the Earth, which passes through the terrestrial poles and through an arbitrary point, namely Greenwich Observatory, so Right Ascension in the sky is measured from a meridian passing through the poles of the celestial equator and a point, the vernal equinox, the point where the ecliptic cuts the equator and the Sun crosses from south to north.

Right Ascension is commonly expressed in hours, minutes, and seconds of time, because it is measured as the time which elapses between the passage over the meridian of any place of

2 POPULAR GUIDE TO THE HEAVENS.

the first point of Aries and of the object whose position is to be defined. Declination is expressed in degrees of arc, because it is usually measured by graduated circles so divided on the instrument.

For further account of the subject of the measurement of positions on the celestial sphere, the student must be referred to the text books of spherical astronomy.

THE HORIZON AND THE ZENITH.

An observer upon the Earth is debarred by the Earth itself from seeing more than one half the sky at once. The plane which touches the Earth at the point where he stands, produced all ways to meet the celestial sphere, cuts it in the great circle which is his celestial horizon. When the observer is looking over the sea, his visible horizon is, owing to the spherical shape of the Earth, depressed below the above defined celestial horizon by an amount which depends on the height of the observer above sea level. This depression is called the "dip of the horizon," and must be taken into account when altitudes of a heavenly body are measured from the visible sea horizon. In other words, the celestial horizon is 90° from the zenith, the point vertically above the observer ; the visible sea horizon is more than 90°, and the observer can by this amount see more than half the sky.

REFRACTION.

The effect of the layers of air through which light must pass on its way from a star to the observer upon the Earth is to raise the star apparently above its real position. The effect of refraction is sometimes very obvious when either the Sun or Moon is close to the horizon, in a flattening of the solar or lunar disc. The closer a body to the horizon, the more it is raised by refraction ; the lower limb of the Sun is consequently raised more than the upper, with the result that the Sun appears no longer round, but flattened.

On the horizon a body is apparently raised 34′ of an arc.

At an elevation of 1°   ,,   ,, $24\frac{1}{3}$′   ,,
    ,,   3°   ,,   ,, $14\frac{1}{3}$′   ,,
    ,,   5°   ,,   ,,   10′   ,,
    ,,   10°   ,,   ,,   5′   ,,
    ,,   20°   ,,   ,, $2\frac{1}{2}$′   ,,
    ,,   30°   ,,   ,, $1\frac{1}{2}$′   ,,
    ,,   40°   ,,   ,,   1′   ,,

Above 40° of elevation the refraction is measured by seconds of arc alone, and becomes less and less till it vanishes at the zenith.

DIURNAL PARALLAX.

It is clear that when a celestial body is viewed from a point on the Earth's surface, it cannot appear precisely in the same direction as when it is viewed from the centre. The difference is named Diurnal Parallax. Owing to the small size of the Earth, and the vast distance of the stars, the Diurnal Parallax of stars is absolutely insensible. With the Sun and planets, however, the case is different. The nearer the body to the Earth, and the nearer to the horizon of the observer, the greater the effect. Upon the Moon the effect is quite large : when the Moon is just rising it appears lower among the stars than it would do from the centre of the Earth, by nearly a degree, or about two diameters of the Moon.

Owing to this large displacement in the position of the Moon as seen from different parts of the Earth, a star which is occulted by the Moon at one place may be completely clear of it at another, and, similarly, while at one place a total eclipse of the Sun is seen, at another the eclipse may not even be partial.

ANNUAL PARALLAX.

It is equally clear from the figure that a star will not be seen exactly in the same direction at different times of year. By far the greater number of stars are so far away that not even the displacement of the Earth by 186,000,000 miles produces any sensible effect. But upon a certain number of nearer stars the effect is just measurable. They shift their positions relative to the more distant stars very slightly during the year ; and this effect is called Annual Parallax.

APPARENT DAILY ROTATION OF THE HEAVENS: RISING AND SETTING.

In consequence of the actual rotation of the Earth, from W. to E., an observer upon it sees the heavens apparently rotating about an axis directed to the pole of the sky. Every celestial body therefore apparently describes once a day a circle round the pole. The pole is elevated above the horizon by an amount equal to the latitude of the observer. Suppose this is 50°. A star nearer the pole than 50° describes the whole of its circle above the horizon ; it never sets, and is called circumpolar. A star further than 50° from the pole will not be circumpolar ; part of its daily circle will be below the horizon, and a greater part the farther the star is from the pole, until for a star 180°—50°, i.e., 130° from the pole, the whole circle is permanently below the horizon ; and in lat. 50° N. this star will never rise at all.

The figure in Plate 2, "apparent diurnal paths of the Sun," illustrates this principle in the case of the Sun, and shows why the days are longer in summer than in winter : the sun is nearer the north pole of the sky in summer than in winter, for we have seen that the ecliptic cuts the equator at a considerable angle (23° 27').

Tides.—The figure in the lower part of Plate 2 illustrates the fact that when a body like the Earth, covered with an ocean, is rotating, the attraction of the Sun or Moon produces disturbances in the level of the ocean, which are called tides. Though these protuberances are caused by the attraction of the tide producing body they are not necessarily, nor indeed generally, in line with it. On an earth whose ocean is much broken up by continents the tidal waves are much broken up and disturbed, and the theory of the tides becomes exceedingly complex. Owing to its relative nearness more than counter-balancing its smallness, the Moon is a much more efficient tide producing agent than the Sun.

The figure at the top of the Plate is to explain the expressions *spring* and *neap* tides. At New Moon and Full Moon the tides raised separately by the Sun and Moon conspire, and an exceptionally high tide is produced, which is called a *spring* tide. At first or last quarter of the Moon, the Sun tends to produce low water when and where the Moon tends to produce high water, and the result is a small or *neap* tide.

THE SIGNS OF THE ZODIAC.

The region of the heavens along the ecliptic, or the *zodiac*, was divided by the ancients into twelve parts, or *signs*, each 30° in length, which took their names from the principal constellations along the zodiac. Thus, starting from the Vernal Equinox, the first sign was called Aries, the second Taurus, and so on. The gradual change in the position of the Equator and Equinox, due to precession, has thrown back the Equinox into the constellation Pisces, and displaced all the signs of the zodiac from the constellations whose names they bear. Nevertheless the old names are retained, and the student must be warned against possible confusion. When the Almanac says "Sun enters Aries; spring commences," a reference to the star maps will show that the Sun is still in Pisces, and will be for a month We must distinguish, therefore, the names of the constellations from the same names applied to the signs of the zodiac.

The Signs of the Zodiac, with their symbols, are given in the following table. Counting from the Vernal Equinox we have—

0° to 30°	...	0.	♈ Aries.		180° to 210°	...	VI.	♎ Libra.
30° to 60°	...	I.	♉ Taurus.		210° to 240°	...	VII.	♏ Scorpio.
60° to 90°	...	II.	♊ Gemini.		240° to 270°	...	VIII.	♐ Sagittarius.
90° to 120°	...	III.	♋ Cancer.		270° to 300°	...	IX.	♑ Capricornus.
20° to 150°	...	IV.	♌ Leo.		300° to 330°	...	X.	♒ Aquarius.
50° to 180°	...	V.	♍ Virgo.		330° to 360°	...	XI.	♓ Pisces.

PLATE 3.

THE ORBITS OF THE INNER PLANETS.

In the attempt to represent the orbits of celestial bodies on maps or charts, it must always be remembered that, except in the case of orbits which happen to lie in the same plane it is impossible to depict on any drawing the veritable position of more than one. We are obliged to resort to some process of a more or less artificial character. For instance, we take the plane of the Ecliptic, that is, the Earth's orbit, as the plane of the paper, and then we simply lay down on it the orbits of the other bodies, notwithstanding that their planes are inclined to the Ecliptic. The points in which the real orbit passes through the plane of representation are called the *Nodes*, the ascending node being that at which the planet passes from the southerly to the northerly side of the plane. Each orbit may be conceived to be turned around its line of nodes till its plane coincides with the Ecliptic. It is thus tha Plate III. is produced.

The path which every planet describes is an *ellipse*, and the Sun is situated in one of the two *foci* of the ellipse, S or H. The longest diameter of the ellipse, P A, which passes through the two foci, is called the *major axis;* the diameter X Y, at right angles to it, is the *minor axis*, and the two intersect in the centre, O, of the ellipse. The points A and P are the *Apsides* of the ellipse. We will suppose the Sun is in the focus

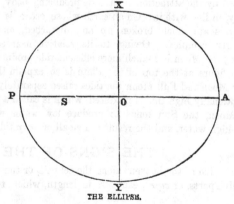

THE ELLIPSE.

S in the figure, and that a planet is describing, under its attraction, the ellipse. The ellipse is called its *orbit*. The point P, nearest to the Sun, is the *Perihelion* of the orbit; the point A, farthest away, is the *Aphelion*. The half major axis, O P, which is also equal to the distance S X, is called the *mean distance*. The ratio of O S to O P is the *eccentricity* of the orbit; the smaller this ratio is, the more does the ellipse resemble a circle. The orbits of the more important planets all have small eccentricities.

It will be convenient to give here the symbols which are in use for the Sun, Moon, and Planets, and certain other signs used occasionally in this work.

Explanation of Astronomical Symbols and Abbreviations.

☉	The Sun.	♂	Mars.	☌	Conjunction.
☾	The Moon.	♃	Jupiter.	☐	Quadrature.
☿	Mercury.	♄	Saturn.	☍	Opposition.
♀	Venus.	♅	Uranus.	☊	Ascending Node.
⊕ or ♁	The Earth.	♆	Neptune.	☋	Descending Node.
h	Hours.	°	Degrees.	N. North.	S. South.
m	Minutes of Time.	′	Minutes of Arc.		
s	Seconds of Time.	″	Seconds of Arc.	E. East.	W. West.

The orbits of the planets Mercury, Venus, Earth, Eros, and Mars are represented in this Plate, and for illustrating the use of it we take the orbit of Mercury. The point A is the Aphelion where the planet is most distant from the Sun. The next point marked is the Ascending Node ☊, where the orbit comes through the plane of the paper, the inclination being 7° 0′, as given in the table in the upper right hand corner of the map. P is the Perihelion, where Mercury is nearest the Sun. For a complete revolution this planet requires a period of 87·969 days. Similar remarks apply to the other orbits. Thus, for instance Mars, the outermost of the four planets shown in this figure, revolves in the period of 686·951 days. Its Perihelion is marked P, and Aphelion A, the Ascending Node is ☊, and the inclination is 1° 51′. The inclinations of the cometary orbits are given in the right hand lower corner of the Plate. The orbits of the three following comets are drawn, Biela's Comet, Comet I. 1866, and Comet III, 1862. These have been chosen because they possess the additional interest of being the paths of the three chief meteor swarms. The famous showers of "Leonids," which used to appear about November 13th, in magnificent displays every 33 years, move in the track of Comet I. 1866. The "Andromedids," or meteors of November 27th, have the same orbit as Biela's Comet, and the "Perseids" pursue the course of Comet III. 1862. In the case of each of the cometary orbits the Descending Node has been marked on the Plate, as it is at this Node that the Earth meets the associated meteor swarm.

The planet Eros is a recently discovered small planet, whose orbit is of a remarkable character. It is very eccentric, and considerably inclined to the ecliptic; and on favourable occasions the planet may be within about 13,000,000 miles of the Earth, nearer to us than any celestial body except our own Satellite.

PLATE 4.

THE ORBITS OF THE OUTER PLANETS.

The innermost orbit on this Plate is that of Mars, for those belonging to planets still closer to the Sun would be too small to be shown in a figure of the scale necessary for the outer planets.

Next to Mars comes the zone of minor planets, of which more than 500 are now known, while some twenty or thirty new ones are discovered by photography yearly. Their orbits are tangled in a way impossible to represent otherwise than conventionally in a figure of small size. They exhibit great diversity of eccentricity and inclination to the ecliptic. And the task of keeping up the computation of the places of this fast growing family of tiny planets is becoming almost beyond reasonable possibility. In our figure the innermost represented is Medusa, with a period of 3·12 years ; the outermost Hilda, with a period of 7·90. A later discovery, Adalberta, lies closer to the Sun than Medusa, with a period of 3·01 years ; and Thule, with a period of 8·86 years, lies considerably farther out than Hilda.

Beyond the zone of minor planets lie the major planets Jupiter, Saturn, Uranus, and Neptune.

On this Plate are also represented the orbits of several interesting comets belonging to the Solar System. The Comet of Encke has the smallest orbit of any known comet, and it is especially interesting because of the somewhat irregular and unexplained acceleration to which it is subject. There is room on this Plate to show the orbit of Biela's Comet, part of which was shown on the preceeding Plate.

Halley's Comet is the only periodic comet which makes a splendid appearance ; the others are all small, many of them almost insignificant objects. But Halley's Comet on its last return had a nucleus as bright as a first magnitude star, and a tail twenty-five degrees long ; and its next return in 1910 will be awaited with very great interest.

A portion of the orbit of the Comet of 1882 is shown on account of its remarkable nature. The comet passed so close to the Sun that it almost grazed, and it swung round 180° of its orbit in the space of three hours.

A remarkable relation, hitherto unexplained, connects the distances of the various planets from the Sun. If we write down the series of numbers—

$$0 \quad 3 \quad 6 \quad 12 \quad 24 \quad 48 \quad 96 \quad 192 \quad 384,$$

and add 4 to each, we have

$$4 \quad 7 \quad 10 \quad 16 \quad 28 \quad 52 \quad 100 \quad 196 \quad 388.$$

The first four are very nearly in the proportion of the distances from the Sun of Mercury, Venus, the Earth, and Mars ; 52 and 100 represent equally well the distances of Jupiter and Saturn ; the intermediate No. 28, which stands for the average minor planet, actually suggested the search for them ; when Uranus was discovered, it was found to fit 196 ; and when there was a suspicion of a planet beyond Uranus, it was assumed that its distance would be nearly represented by 388. But it is not ; 300 is the real number, and here the rule, which is called Bode's law, breaks down completely.

Plate 1

THE CELESTIAL SPHERE

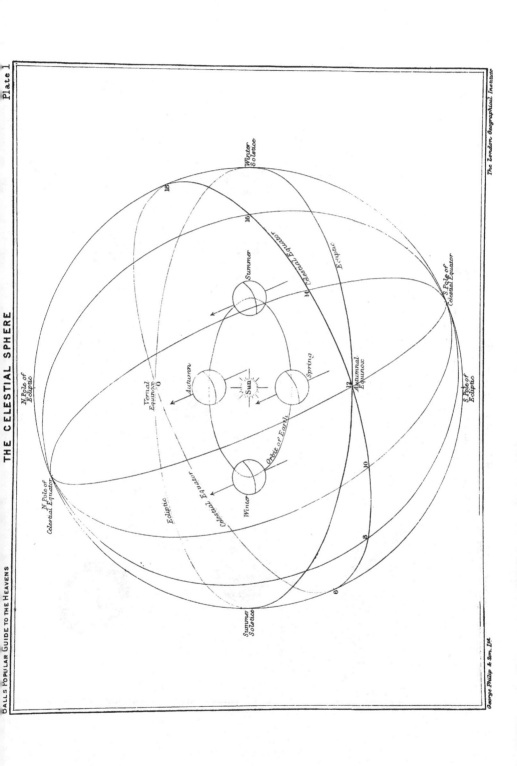

George Philip & Son, Ltd.

The London Geographical Institute

Plate 2

THE SEASONS AND THE TIDES.

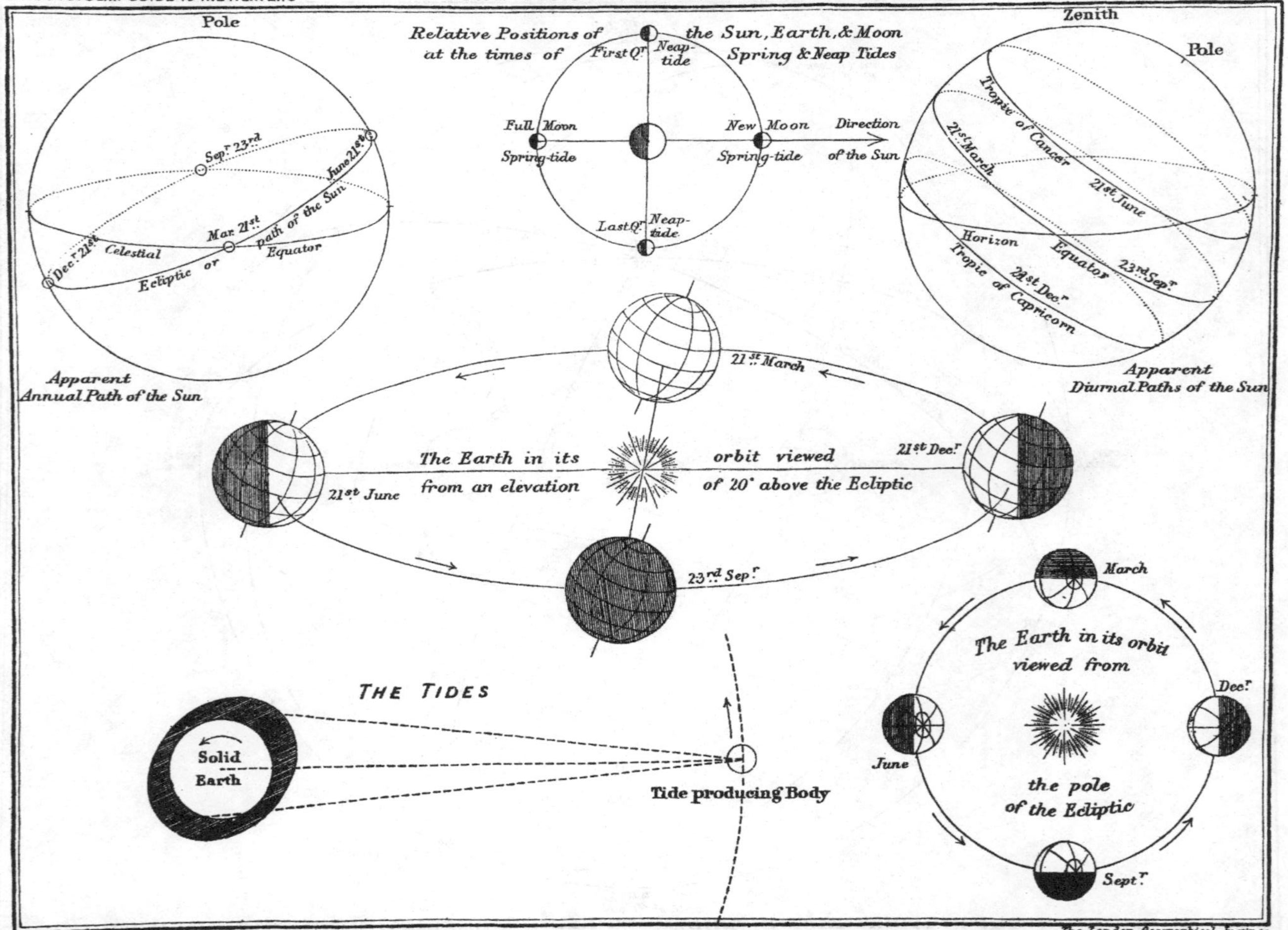

Pole

Zenith

Tropic of Cancer   21st June

21st March   Horizon   Equator   23rd Sep.

Tropic of Capricorn   23rd Dec.

Apparent Diurnal Paths of the Sun.

Relative Positions of the Sun, Earth, & Moon at the times of Spring & Neap Tides

First Qr   Neap-tide

New Moon   Direction of the Sun.

Spring-tide

Full Moon   Spring-tide

Last Qr   Neap-tide

21st March

21st Dec.

orbit viewed of 20° above the Ecliptic

23rd Sep.

The Earth in its orbit viewed from an elevation

21st June

The Earth in its orbit viewed from the pole of the Ecliptic

Dec.

March

June

Sept.

Pole

East 21st June

Sep. 23rd   Equator

Mar 21st   Celestial or   Ecliptic

Dec 21st   path of the Sun

Apparent Annual Path of the Sun.

THE TIDES

Tide producing Body

Solid Earth

George Philip & Son, Ltd.

The London Geographical Institute

Plate 3

BALL'S POPULAR GUIDE TO THE HEAVENS

THE INNER PLANETS.

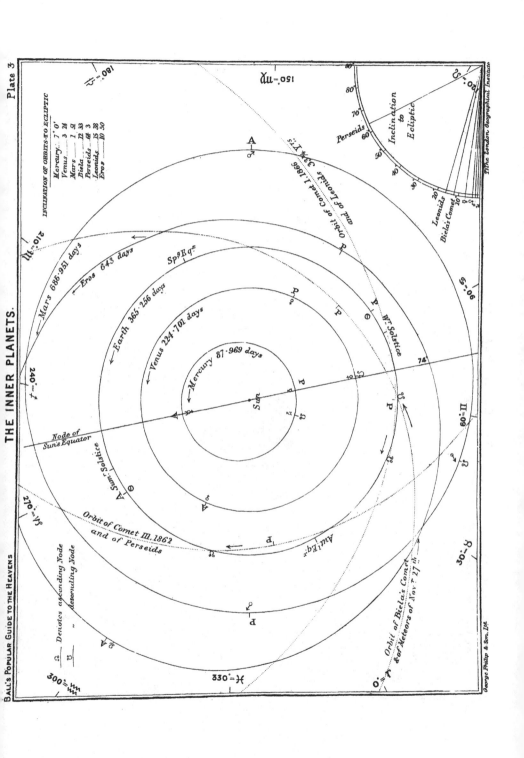

INCLINATION OF ORBITS TO ECLIPTIC

Mercury	7° 0'
Venus	3 24
Mars	1 51
Biela	12 33
Perseids	64 3
Leonids	15 38
Eros	10 50

Mars 686·951 days

Eros 643 days

Earth 365·256 days

Venus 224·701 days

Mercury 87·969 days

Sun

Sp.g Eq.ʳ

W. Solstice

Node of Sun's Equator

Sun. Solstice

Orbit of Comet III. 1862 and of Perseids

Orbit of Comet I. 1866 and of Leonids 33¹⁄₄ yᵣˢ.

A·witir·qⁱᵘ

Orbit of Biela's Comet & of Meteors of Nov. 27 th.

Inclination to Ecliptic

Perseids

Leonids

Biela's Comet

Eros

Denotes ascending Node
" descending Node

The London Geographical Institute.

George Philip & Son, Lᵈ.

180°= ♎︎
150°= ♍︎
210°= ♏︎
120°= ♌︎
240°= ♐︎
90°= ♋︎
270°= ♑︎
60°= ♊︎
300°= ♒︎
30°= ♉︎
330°= ♓︎
0°= ♈︎

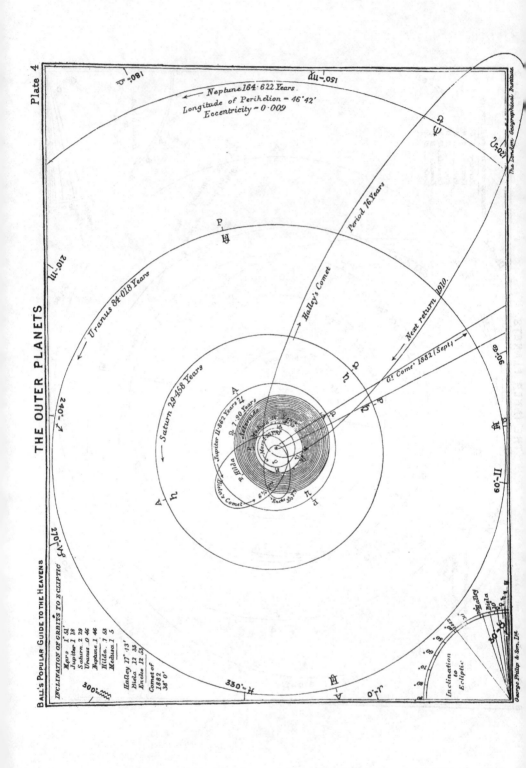

Plate 4

THE OUTER PLANETS

BALL'S POPULAR GUIDE TO THE HEAVENS

Neptune 164·622 Years
Longitude of Perihelion = 46°42'
Eccentricity = 0·009

Uranus 84·018 Years

Saturn 29·458 Years

Halley's Comet
Period 76 Years
Next return 1910.

G: Comet 1882 (Sept.)

Jupiter 11·862 Years

Asteroides

INCLINATION OF ORBITS TO ECLIPTIC

Mars	1° 51'
Jupiter	1 18
Saturn	2 29
Uranus	0 46
Neptune	1 46
Hilda	7 63
Medusa	1 5

Halley	17° ·13'
Biela	12 33
Encke	12 53
Comet of 1882	38° 0'

Inclination to Ecliptic

The London Geographical Institute

George Philip & Son, Ltd.

CHAPTER II.

PLATE 5.

THE SIZE OF THE PLANETS.

Plate 5 has been drawn to show the relation and actual sizes of the different planets and of the rings of Saturn. The determination of the size of a planet in miles is by no means a simple matter. Firstly, it is necessary to measure the angular size of the planets, as seen from the Earth, in seconds of arc, and this is affected by the phenomenon of irradiation by which a bright object against a dark background looks larger than it really is. Also the planets Uranus and Neptune are so far away and look so small that it is hard to measure them, and it is still quite uncertain which is really the larger. We have therefore shown them equal. When the angular diameters at a given distance have been found, we next require a knowledge of the Solar Parallax, that is, of the angular radius of the Earth as seen from the Sun ; the adopted value of this quantity is 8″80. And, finally, we require the diameter of the Earth in miles ; for this we adopt Col. Clarke's latest value, 7926·6 miles.

We can then calculate the diameters of the planets given below, and shown to scale upon the Plate. The times of axial rotation and the inclination of the planets' equator to the Ecliptic are given when they are known.

		Diameter in miles.		Time of Rotation.		Inclination of Equator to Ecliptic.
Mercury	... ...	3,000	...	Unknown	...	Unknown
Venus	... ...	7,700	...	Unknown	...	Unknown
Earth	... ...	7,926	...	23ʰ· 56ᵐ· 4ˢ·	...	23° 27′
Mars	... ...	4,300	...	24 37 23	...	26 21
Jupiter	{ Equatorial	90,000	...	9 55	...	3 5
	{ Polar	84,000				
Saturn	{ Equatorial	76,000	...	10 14 0	...	28 10
	{ Polar	70,000				
Uranus	... ...	31,000	...	Unknown	...	Unknown
Neptune	... ...	31,000	...	Unknown	...	Unknown

Dimensions of Saturn's rings :

Outer radius of outer ring	...	...	...	86,000 miles.
Inner radius of outer ring	...	...	...	75,000 „
Outer radius of inner ring	...	...	...	73,000 „
Inner radius of inner ring	...	...	...	55,000 „
Inner radius of dusky ring	...	...	...	44,000 „

All these dimensions are based upon a careful comparison of the most recent measures made with the great telescopes at Lick, Yerkes, Washington, and Greenwich Observatories. The time of rotation and position of the equator of Mercury, Venus, Uranus, and Neptune are unknown because of the want of definite markings on those planets. There is considerable reason to suppose that the periods of rotation of Mercury and Venus are the same as their periods of revolution round the Sun, namely 88·0 and 224·7 days. In that case they would always turn the same face towards the Sun. If we assume, which is probably true, that the satellites of Uranus and Neptune move nearly in the planes of the planets' equators, then we may add to the last column for those planets, 101° and 145°, the inclinations being given greater than 90° to conform to the fact that the satellites, unlike any other bodies in the Solar System, move in a retrograde direction.

Recent measures have given the following diameters for the four principal minor planets :—

Ceres	...	...	...	480	miles.
Pallas	...	...	...	300	,,
Juno	...	...	...	120	,,
Vesta	...	...	...	240	,,

PLATE 6.

PHASES OF THE PLANETS AND OF SATURN'S RINGS.

This Plate exhibits the appearances of the planets Mars, Venus, and Saturn when occupying different parts of their orbits. A reference to Plate 3 makes it clear that the distance between the Earth and Mars must vary considerably at different dates, according to the positions which the bodies occupy in their paths around the Sun. Of course, if the orbits were both circular, it is clear that the greatest possible separation between the two bodies would be attained at every *conjunction ;* that is to say, whenever the Earth, Sun, and Planet are in a straight line (at least, in their projected orbits), the Earth and Planet being at opposite sides of the Sun. The same diagram makes it plain that the least distance apart would occur at every *opposition;* that is, whenever the three bodies, as represented in their projected orbits, were in a straight line, with the Earth in the middle.

The eccentricity of the orbit of Mars considerably modifies the circumstances. It will ·be seen, by referring to Plate 3, that an opposition occurring in the latter half of the year will generally be more favourable (*i.e.*, bring the two bodies closer together) than one in the first half of the year, and that the most favourable opposition happens when the Earth and Planet are situated in about 333° longitude   On the other hand, an opposition occurring in longitude 153° will be as unfavourable as possible. The Earth's longitude on August 26th is 333°, and on February 22nd it is 153° ; hence the most favourable opposition of Mars will occur on August 26th, and the closer to that date the opposition happens the better. The most unsuitable oppositions are about February 22nd.

The greatest distance at which the two planets can possibly be separated is attained when the Earth's longitude is 333°, and that of Mars 153° ; that is to say, when *conjunction* occurs about August 26th.

Figures 1, 4, and 5, in the upper part of the left-hand portion of Plate 6, show the relative apparent sizes of the planet—at most favourable opposition (August 26th), at least favourable opposition (February 22nd), and at its greatest possible distance. These views illustrate the advantage of an opposition occurring somewhere near the end of August, when the appearance of the planet is to be studied.

When the lines from the Sun to the Earth and the Sun to the Planet are at right angles, the Planet is said to be in *quadrature*. A very distinct phase is then perceptible in Mars, by which about a quarter of its diameter is cut off. The appearances of the planet at western and eastern quadrature, as shown in an inverting telescope, and the apparent size of the planet on the same scale as the other figures, is also given. For the topography of the planet, the reader may refer to Plate 8. As to the times and seasons for observing Mars in its vary ing aspects, reference may be made to the Index to Planets, *see* pages 38· and 42.

Since the orbit of Venus lies inside that of the Earth, the appearances of this planet differ considerably from those of an exterior planet like Mars. It is obvious that the nearest approach of the two bodies will occur at *inferior conjunction*, or when Venus and the Earth are on the same side of the Sun; and that the greatest distance between them will occur at *superior conjunction*, or when the two bodies are at opposite sides of the Sun. It might, at first sight, therefore, be supposed that at inferior conjunction the planet would be seen best, being then apparently largest ; and that it would be least favourably placed at superior conjunction. The relative apparent sizes of this planet *just before* inferior, and at superior, conjunction are shown in the lower part of the left-hand portion of this plate ; but since, in the former configuration, the illuminated part of the globe is reduced to a very thin crescent, and since in both cases the planet ls enveloped in the Sun's rays, in neither of these phases is it suitably situated for observation.

Venus attains its greatest brightness as an evening star about a month after its greatest elongation east. The greatest brightness of the same planet as a morning star precedes by about a month its greatest elongation west.

The second figure has been drawn to represent the size and shape of Venus when most brilliant. The third figure exhibits the appearance of Venus when situated at a distance of 40° from the Sun in the further part of its orbit. In this position it presents a *gibbous* form. It will be seen, however, that the diminution of light caused by its increased distance from the Earth, more than compensates for the larger proportion of the illuminated surface visible, so that, on the whole, the amount of light received from the planet is less than when it is in the position corresponding to Figure 2. In the Index to Planets, p. 39, the method of finding the position of Venus for any date up to 1950 is explained.

For the general details of the planet Saturn reference may be made to Plate 11. In this place we discuss only the varying appearances of the rings. The right-hand portion of Plate 6 contains twelve figures depicting the different aspects which the ringed planet presents according to the position it happens to occupy in its orbit. In connection with the Table of Planetary Phenomena, p. 38, this plate will enable the reader to determine with considerable accuracy the appearance of the rings at any time. If the opposition of Saturn occurs in the middle cf January in any year, it will be found that Fig. 1 represents the system. The rings are then opened nearly to their full extent, and the upper portion of the ball just extends beyond the outer margin of the rings. If the opposition occurs in February, the rings will be found to have closed up somewhat, and to appear as shown in Fig. 2. If the opposition occurs in March, the rings will shrink almost to a straight line, as in Fig. 3. At oppositions occurring in April, May, and June, the appearances will be as in Figs. 4, 5, and 6, the rings appearing the more open the more nearly the date of opposition approaches June. Figs. 7—12, in a similar way, show the changes which this system will undergo at oppositions occurring in the latter six months of the year.

It must, of course, be understood that the appearance here depicted for any month will not recur every year in that month, but will only be seen in those years in which the opposition of the planet occurs during the month in question, and then only with accuracy at the date of opposition. But, as Saturn takes a period of no less than 29½ years to accomplish its revolution, the alteration in its appearance will vary very little for several months before and after opposition, so that the figure for any month may be taken to represent the appearance of the system during the year in which opposition occurs in that month. Thus, in the year 1921, the Table of Planetary Phenomena tells us that the opposition of Saturn takes place in March, whence we learn that during this year the rings will be almost edgewise towards us. Again, in the year 1928, opposition occurs in June, from which we infer that during that year the rings will be open to their fullest extent, and most favourably situated for observations

These pictures have, as usual, been drawn to represent the planet as seen in an astronomical telescope, which always inverts the object, so that Figs. 3—8 exhibit the appearance of the system when the northern face of the ring is tilted towards us so as to become visible, while in Figs. 1 and 2, and 9—12, it is the southern side of the rings which is seen.

To facilitate reference, a column has been added to the Table of Planetary Phenomena, p. 38, to show which of the phases are presented in the corresponding opposition. For example, if the opposition is in October, the column alluded to gives the number 10, which means that during the year in question the planet Saturn will present, when visible at all, a phase resembling that shown in Fig. 10 on Plate 8.

At the times when the ring is seen edgewise the sequence of appearances may be very complicated. The ring may become quite or very nearly invisible from any one of three causes ;

(1.) The plane of the ring may pass through the Earth, in which case the ring is seen exactly edgewise. And as the ring is very thin, its illuminated edge is not bright enough to be seen, and the ring disappears completely in all telescopes.

(2.) The plane of the ring may pass between the Earth and the Sun, in which case the Sun is shining on the opposite side of the ring to that which is presented, very obliquely, to the Earth ; and in this case the ring is almost invisible, even in great telescopes.

(3.) The plane of the ring may pass through the Sun, in which case neither side of the ring is effectively illuminated, and again it almost disappears.

It is sometimes a matter of discussion how big a planet looks in a telescope of a given magnifying power. By means of this Plate the question may be answered. The Plate is drawn to such a scale that if it is placed ten feet from the eye the figures of the planetary discs subtend the same angle as the planets' images themselves, at the corresponding phases would do when seen in a telescope which gives a magnifying power of 80 diameters.

PLATE 7.

SYSTEMS OF SATELLITES.

This Plate exhibits the relative dimensions of the orbits of the systems of satellites attending certain of the planets. With the exception of the system surrounding Mars, which is on a scale twenty times as large as the rest, the orbits are all laid down on a uniform scale of half a million miles to the inch. The periods of revolution of the satellites around their primaries are also marked on the orbits approximately. More complete numerical information than it has been found convenient to represent on the map is given in the following tables.

THE SATELLITE OF THE EARTH : THE MOON.
    Mean distance from the centre of the Earth : 239,000 miles.
    Periodic time : 27 days, 7 hrs. 43 mins. 11 secs.

THE SATELLITES OF MARS :	Mean distance from Centre of Planet.		Periodic Time.			
			Days	hrs.	mins.	secs :
Phobos	5,850 miles	... ...	0	7	39	14
Deimos	14,650 „	... ...	1	6	17	55
THE SATELLITES OF JUPITER :						
V. (Nameless) ...	112,500 „	... ...	0	11	57	23
I. (Io)	261,000 „	... ...	1	18	27	34
II. (Europa)	415,000 „	... ...	3	13	13	42
III. (Ganymede) ...	664,000 „	... ...	7	3	42	33
IV. (Callisto)... ...	1,167,000 „	... ...	16	16	32	11

The Satellites of Saturn :

Mimas...	...	...	117,000	„	...	...	0	22	37	5
Enceladus	...	...	150,000	„	...	...	1	8	53	7
Tethys...	...	...	186,000	„	...	...	1	21	18	26
Dione ...	...	...	238,000	„	...	...	2	17	41	10
Rhea ...	...	...	332,000	„	...	...	4	12	25	12
Titan ...	...	...	771,000	„	...	...	15	22	41	27
Hyperion	...	...	934,000	„	...	...	21	6	38	24
Iapetus	...	...	2,225,000	„	...	...	79	7	56	23

In July, 1904, Prof. E. C. Pickering announced the confirmation of the discovery of a ninth satellite of Saturn, Phœbe, first found on photographs in 1899. Its period is about 1½ years, and its distance from Saturn about 8,000,000 miles.

The Satellites of Uranus :

			Mean distance from Centre of Planet.				Days	Periodic Time. hrs.	mins.	secs.
Ariel ...	...	...	120,000 miles	...	...		2	12	29	21
Umbriel	...	...	167,000 „	...	...		4	3	27	37
Titania	...	...	273,000 „	...	...		8	16	56	30
Oberon	...	...	365,000 „	...	...		13	11	7	6

The Satellite of Neptune :

| (Nameless) | ... | ... | 221,500 | „ | ... | ... | 5 | 21 | 2 | 38 |

The satellites of Mars are a remarkable pair. The inner, Phobos, is the only satellite that revolves faster than its primary rotates. In consequence of this, it must rise in the west, and set in the east. The outer satellite, Deimos, revolves in a period so little greater than that of the planet that it goes through all its phases twice between the times of rising and setting.

The four well known satellites of Jupiter are almost always called by their numbers ; their names, which have almost fallen into disuse, are therefore placed in brackets. The fifth satellite, discovered by Barnard in 1892, still remains without a name.

Recent measures have given the following values for the diameter of the four large satellites of Jupiter, and of Titan, the largest of the satellites of Saturn.

I.	2,500 miles.	III.	3,600 miles.
II.	2,200 „	IV.	3,300 „
	Titan	...	2,900 miles.

But it is very probable that, on account of the effect of irradiation, these diameters may be several hundred miles too large.

The diameters of the other satellites of our system are, for the present, beyond the reach of measurement. If we measure the amount of light they reflect from the sun, and make some assumption as to their *albedo*, or light reflecting power, we can estimate roughly the probable diameters of the others. One finds that the Phobos and Deimos are probably about 10 and 30 miles in diameter, the fifth satellite of Jupiter, 100.

PLATE 8.

MAP OF MARS.

The selection for this work of a representation of the surface of the planet Mars is a matter of great difficulty. Observers of Mars are divided into two camps—those who see the canals and those who do not. The former are in the strong position that they are perfectly sure that they see what they represent in their drawings ; the latter declare that under the finest possible conditions of observation, and with the most perfect instruments, they can see nothing resembling the straight markings which are known as canals. And further, they bring forward experiments which make it clear that irregularly disposed markings imperfectly seen, give the effect of straight streaks, by an optical illusion. The interest which has been excited by the speculations based upon the drawings of these apparently artificial markings makes it impossible to present a chart of Mars in which the canals are omitted. We give, therefore, a reproduction of the chart of Mars made at the Lowell Observatory, Flagstaff, Arizona, by Mr. A. E. Douglass, from a study of all the drawings made there by various observers during the opposition of 1896-97. At the same time it is necessary to give the caution that some of the very best observers deny altogether the truth of this representation of the planet.

Our difficulty is increased by the fact that there are two rival systems of nomenclature for the features of Mars—an earlier system in which the so-called lands and seas are named after modern Astronomers—Herschel, Leverrier, Dawes, &c., and a later, in which the names are taken from classical geography and mythology. The later system seems likely to prevail, and we have adopted it in the present work. It is useless to give a catalogue of some 400 names of markings whose very existence is in dispute. We confine ourselves therefore to naming some of the more prominent features, to which a number is affixed in the plate.

1. Fastigium Aryn.		13. Mare Tyrrhenium.
2. Margaritifer Sinus.		14. Syrtis Minor.
3. Mare Erythraeum.		15. Syrtis Major.
4. Aurorae Sinus.		16. Cerberus (Canal).
5. Ganges (Canal).		17. Mare Icarium.
6. Lunae lacus.		18. Edom promontorium.
7. Solis lacus.		19. Hellas.
8. Sirenius lacus.		20. Ausonia.
9. Mare Sirenum.		21. Trivium Charontis.
10. Eumenides (Canal).		22. Orcus (Canal).
11. Mare Cimmerium.		23. Pyriphlegethon.
12. Charontis lacus.		24. Mare Chronium.

It should be understood that in the unsteady air of England it is almost hopeless to expect to see many of the finer details. Not even in the most favourable climates are they to be seen always, or all at once. And much training of the eye is required before it is fit to decide for or against the existence of these details on the very verge of invisibility.

<hr>

PLATE 9.

JUPITER AND SATELLITE I.

Owing to the absence of permanent features on Jupiter it is not possible to give a map of the planet. From year to year the position and breadth of the belts change, the tints of the surface change, and the shape and character of the spots change. Under these circumstances the best that can be done is to present drawings of the planet which are typical, yet possess features of more than average interest. We therefore select a set of drawings covering the period when the "great red spot" was most conspicuous. It was first seen in July 1878, and in the following year it was the most conspicuous feature on the planet (Figs. 4, 5, 6). In 1880 and 1881 it changed but little (Figs. 7, 8, 9, 12), but after that began to fade ; and at the present time it is visible only as an indentation or scar on the southern equatorial belt. It was evident almost from the first that its period of rotation was not the same as that of the average spot in the belt near it. These gained 22 secs. upon it at each rotation. And though the spot is more or less permanent its own time of rotation has changed by 6ˢ, and for these facts no satisfactory theory has been suggested.

The satellites of Jupiter were the first discoveries made with the telescope, and they remain the most beautiful and interesting objects that a small telescope can show. Their eclipses and occultations and transits over the planet's disc are predicted in the *Nautical Almanac* year by year, to which reference may be made also for the configuration of the satellites each night.

With a very powerful telescope the phenomenon of the transit of satellite I. is very curiously varied. The figures are from drawings made by Prof. Barnard at the Lick Observatory. It had been noticed that when Satellite I. was crossing the disc of the planet, is sometimes appeared double and sometimes very elongated. The drawings supply the explanation. The satellite, not unlike its primary, has a bright equatorial region and darker poles. When it is projected upon a dark belt of a planet the former alone is seen ; when upon a bright belt the latter. The drawing made November 19, 1893, shows the phenomena most completely. The satellite was seen against tne boundary separating a bright from a dark belt ; and it was also partly superposed upon its own shadow. It is scarcely necessary to add that, since the whole apparent diameter of the satellite is little more than a second of arc, it requires the finest telescope and skill to see what is here shown.

PLATE 10.

SATURN.

We are indebted again to Professor Barnard and the Lick telescope for the drawing which has been chosen to illustrate the appearance of the planet Saturn. Although spots are sometimes seen upon the planet, they are uncommon, and the surface markings are usually no more than a few vague dusky belts ; the interest lies in the rings.

In looking at the plate we must imagine the sun behind us and a little to the left. The shadow of the ball is seen upon the rings (at the right hand limb) and the shadow of the rings is seen upon the ball (above). The Cassini division was plainly visible all round, but the Encke division in the outer ring was not visible at the time ; it seems to be a thin place in the ring rather than an actual division. The dusky, or crape ring, showed steely blue against the sky, and at its inner edge was so transparent that the planet could be seen through it. Where it joins the inner ring there is no division, but the two rings merge rapidly the one into the other. The brightest part of the whole is the outer edge of the inner bright ring.

Since this drawing was made the rings have opened out to their fullest extent, and are now (1903) closing in again as the planet approaches the interesting point in its orbit where the rings are seen edgewise.

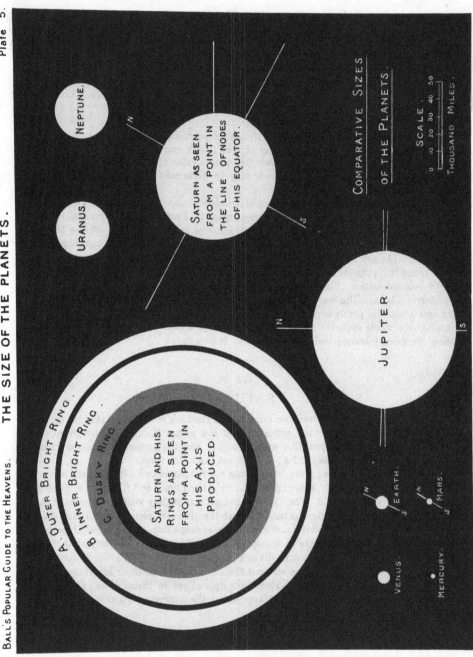

A. OUTER BRIGHT RING.

B. INNER BRIGHT RING.

C. DUSKY RING.

SATURN AND HIS RINGS AS SEEN FROM A POINT IN HIS AXIS PRODUCED.

NEPTUNE.

URANUS.

N

S

SATURN AS SEEN FROM A POINT IN THE LINE OF NODES OF HIS EQUATOR.

N

S

JUPITER.

N

S

EARTH.

N

S

MARS.

VENUS.

MERCURY.

COMPARATIVE SIZES
OF THE PLANETS.

SCALE.

0   10   20   30   40   50

THOUSAND MILES.

George Philip & Son, Ltd.   The London Geographical Institute.

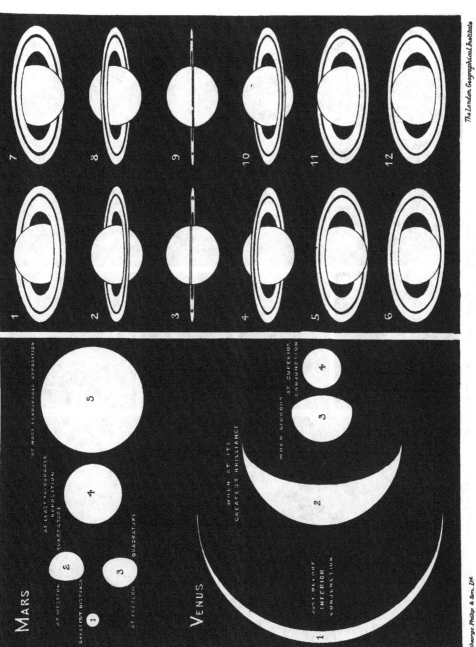

BALL'S POPULAR GUIDE TO THE HEAVENS. PHASES OF THE PLANETS & OF THE RINGS OF SATURN. Plate 6.

MARS

AT WESTERN QUADRATURE

AT GREATEST DISTANCE

AT LEAST FAVOURABLE OPPOSITION

AT MOST FAVOURABLE OPPOSITION

AT EASTERN QUADRATURE

VENUS

JUST BEFORE INFERIOR CONJUNCTION

WHEN AT ITS GREATEST BRILLIANCE

WHEN GIBBOUS AT SUPERIOR CONJUNCTION

George Philip & Son, Ltd.

The London Geographical Institute.

JUPITER.

IV   16ᵈ   16ʰ   32ᵐ

III   7ᵈ   3ʰ   43ᵐ

II   3ᵈ   13ʰ   14ᵐ

I   1ᵈ   18ʰ   28ᵐ

V   11ʰ   57ᵐ

MARS, ON 20 TIMES AS LARGE A SCALE AS THE OTHERS

DEIMOS   1ᵈ   6ʰ   18ᵐ

PHOBOS   7ʰ   39ᵐ

EARTH.

MOON   27ᵈ   7ʰ   43ᵐ

NEPTUNE.

5ᵈ   21ʰ   3ᵐ

URANUS.

ARIEL   2ᵈ   12ʰ

UMBRIEL   4ᵈ   3ʰ   28ᵐ

TITANIA   8ᵈ   16ʰ   56ᵐ

OBERON.

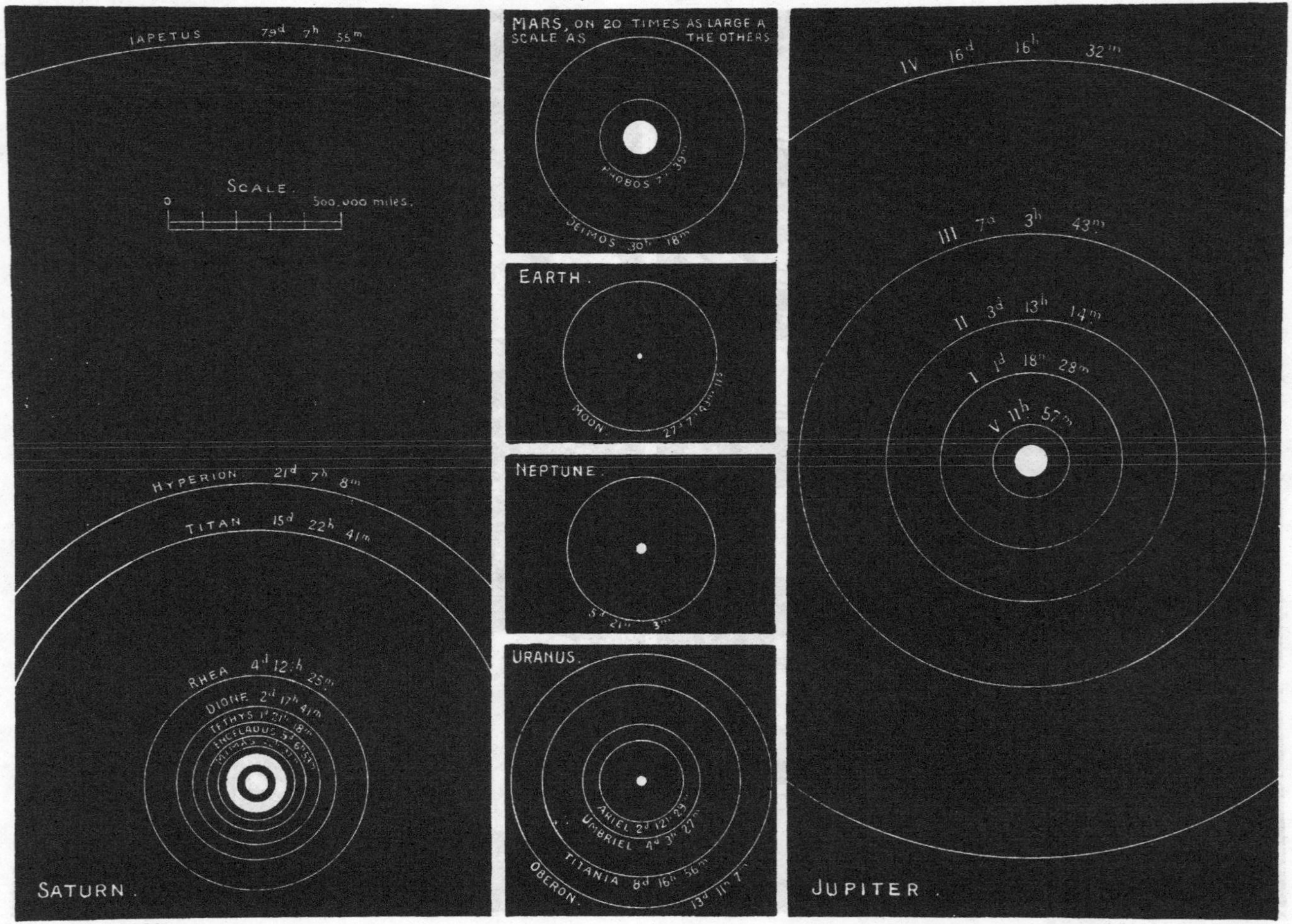

IAPETUS   79ᵈ   7ʰ   55ᵐ

SCALE   500,000 miles.

0

HYPERION   21ᵈ   7ʰ   8ᵐ

TITAN   15ᵈ   22ʰ   41ᵐ

RHEA   4ᵈ   12ʰ   25ᵐ

DIONE   2ᵈ   17ʰ   41ᵐ

TETHYS   1ᵈ   21ʰ   18ᵐ

ENCELADUS   1ᵈ   8ʰ   53ᵐ

SATURN.

Plate 8.

BALL'S POPULAR GUIDE TO THE HEAVENS.   MAP OF MARS, 1896-97.

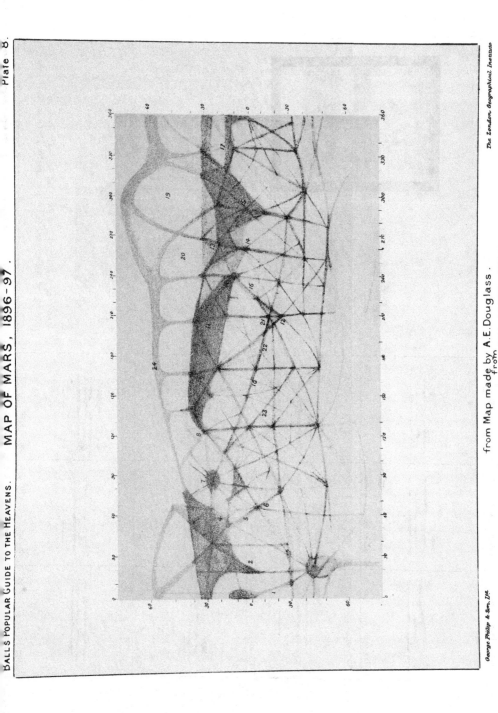

George Philip & Son, Ltd.

The London Geographical Institute.

from Map made by A.E. Douglass.
from
Drawings at the LOWELL OBSERVATORY

Plate 9

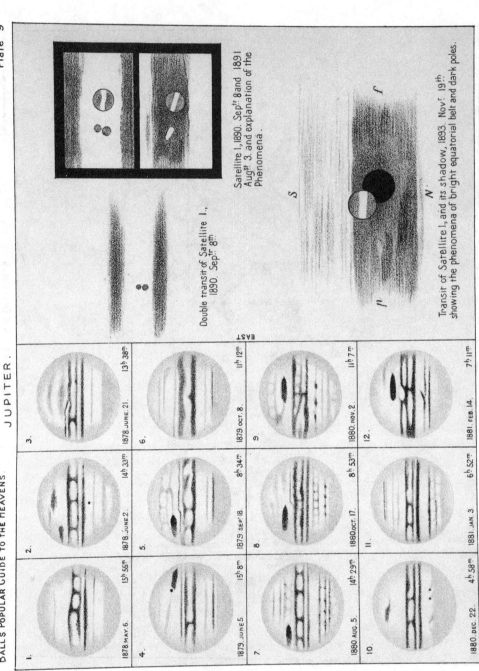

BALL'S POPULAR GUIDE TO THE HEAVENS.

JUPITER.

Double transit of Satellite I., 1890. Sept. 8th.

Satellite I., 1890. Sept. 8 and 1891 Augst 3. and explanation of the Phenomena.

Transit of Satellite I., and its shadow, 1893. Novr 19th. showing the phenomena of bright equatorial belt and dark poles.

EAST

WEST

1. 1878. MAY. 6.   13h 55m
2. 1878. JUNE 2.   14h 33m
3. 1878. JUNE. 21.   13h 38m
4. 1879. JUNE 5.   15h 8m
5. 1879. SEP. 18.   8h 34m
6. 1879. OCT. 8.   11h 12m
7. 1880. AUG. 5.   14h 29m
8. 1880. OCT. 17.   8h 53m
9. 1880. NOV. 2.   11h 7m
10. 1880. DEC. 22.   4h 58m
11. 1881. JAN. 3.   6h 52m
12. 1881. FEB. 14.   7h 11m

George Philip & Son, Ltd.   The London Geographical Institute

SATURN, JULY 2ND 1894.

Plate 10.

BALL'S POPULAR GUIDE TO THE HEAVENS.

The London Geographical Institute

George Philip & Son, Ltd.

CHAPTER III.

PLATES 11 & 12.

THE SUN.

A very important branch of the work of the Royal Observatory, Greenwich, is the daily record, by photography, of the number and size of the spots which appear upon the Sun's surface.   To fill the gaps caused by cloudy weather in England photographs are taken also in India and Mauritius, and are sent home to Greenwich, so that there are very few days in the year for which there is no record.   The purpose of this continuous survey of the Sun is to determine the laws which govern the changes in the area and position of the spots.   It is well-known that the number of spots reaches a maximum about every eleven years ; that at the beginning of each new period the spots are found in higher solar latitudes than at the end ; and that there is an unmistakable, but unexplained, connection between the frequency of Sun spots, of displays of the Aurora Borealis, and of terrestrial magnetic storms.   One of the finest Sun-spot photographs ever taken at Greenwich is reproduced, by permission of the Astronomer Royal, in Plate 11.   The structure of the group is very complex.   Every large spot is accompanied by a crowd of smaller spots, which change comparatively quickly.   A large regular spot consists of two well-defined portions—the black central *umbra* and the surrounding grey *penumbra*.   In the latter, the bright granules which form the *photosphere* of the Sun are elongated and drawn in towards the centre of the spot, making the structure of the photosphere somewhat like thatch.   Very frequently bright bridges are thrown across from one side to the other, and this is generally the prelude to the filling up of the spot.

Sun spots are the seat of tremendous activity in the layers of glowing gas lying above the photosphere.   The most remarkable of the gaseous *prominences*, which stand out above the limb of the Sun when it is totally eclipsed, are almost always associated with spots lying beneath them.   The lower part of Plate 12 is the reproduction of a photograph taken by Prof. Barnard and Mr. Ritchey during the total eclipse of 1900—May 28th.   These prominences are outbursts of hydrogen, calcium, and occasionally of other metallic vapours, which are often thrown up from the surface of the Sun with enormous velocities.   The prominences are conspicuously seen during a total eclipse because the glare in our atmosphere, which ordinarily surrounds the Sun, is then for the moment removed.   The application of a spectroscopic method now enables us to abolish the effect of this glare at any time, and it is possible to make a daily record of the prominences.

The upper part of Plate 12 is a reproduction of a set of drawings of a single prominence made in this manner by Herr Fenyi, at Kalocsa in Hungary, on 1895, July 15.   His description is as follows.   At 7.10 a.m., Greenwich M.T., a very delicately formed prominence stood precisely on the place where a considerable group of Sun spots was passing out of sight round the limb (Fig. 1).   When Fig. 2 was drawn at 7.40, the form of the prominence was changing with extraordinary rapidity.   Determinations of the velocity with which parts of the prominence were moving gave results up to 500 miles per second.   Fig. 3 was drawn at 8.7 ; Fig. 4 at 8.30, when the prominence had reached its greatest height of about 100,000 miles.   At 8.45 its shape had changed very much, and at 9.35, when Fig. 6 was drawn, the great protuberances had completely gone, and the prominence had returned to nearly its appearance of 2½ hours before.

PLATE 13.

PATHS OF SPOTS ACROSS THE SUN'S DISC.

By the rotation of the Sun on its axis, the spots appear to be carried across the disc, along paths parallel to its equator.

The axis, around which the Sun rotates, is inclined to the ecliptic at an angle of 82° 45'. The inclination of the Sun's equatorial plane to the ecliptic is therefore 7° 15'.

The ascending node of the Sun's equator is the point at which a spot on the equator of the Sun would be carried by the Sun's rotation from the southern to the northern side of the ecliptic, and the longitude of the node is the angle which the direction of this point makes with the direction of the *First Point of Aries* as seen from the Sun's centre.   The actual value of the longitude of the ascending node is 74°.   Its position is marked on Plate 2.

Plate 13 shows the paths along which the spots appear to travel at different dates.   They are here represented as actually on the face of the Sun, and not as seen through the inverting telescope that the astronomer ordinarily uses.

On December 6th, the Earth is in the line of nodes, and, consequently, in the plane of the Sun's equator, and the paths pursued by the spots will therefore appear projected into straight lines.   Again, on June 5th, when the Earth is in the opposite point of its orbit, it will be again in the plane of the Sun's equator, and the paths of the spots will again appear projected into straight lines.

On March 4th, the Earth, being then 90° from the node, will be depressed below the Sun's equator by an angle of 7° 15', and the paths of the spots will appear as ellipses of considerable curvature, with their convexities towards the north ; while, on September 6th, from the opposite point of the orbit, the same curves will reappear, only that they will now be convex towards the south.   From March till June, and from September till December, the curvature is decreasing, while in the intervening periods corresponding changes take place in the opposite direction.   We may describe these changes in a somewhat different way by saying that on June 5th and December 6th both poles of the Sun are visible just on the edge of its disc ; from June to December the north pole only is visible ; and from December to June the south pole only can be seen.

By the " position angle of the Sun's axis," is meant the angle which the projection of the northern half of the Sun's axis on its apparent disc makes with the meridian passing through the Sun's centre, reckoned positive towards the eastern, and negative towards the western side of the disc.   If the observation is made at noon, it is the angle which the direction of the axis makes with the vertical, when the image is viewed projected on a sheet of paper placed behind the eyepiece of an inverting telescope.   If the observer's back be turned towards the

Sun, the position angle will be positive when the upper half of the axis leans towards the right, and negative when it leans towards the left. On such a projection the cardinal points, N., S., E., W., lie just as they do in an ordinary terrestrial atlas. On January 5th and July 6th, the position angle of the Sun's axis is Zero ; from July 6th it gradually increases in a positive direction until it reaches its greatest value, viz. : + 26° 20', on October 10th. From this date it gradually diminishes till January 5th, after which it becomes negative, reaching its greatest negative value, viz. : – 26° 20', on April 5th, and returning once more to Zero on July 6th.

PLATE 14.

PHASES OF THE MOON: LUNAR AND SOLAR ECLIPSES.

Phases of the Moon.—The hemisphere of the Moon that is turned towards the Sun is, of course, brilliantly lighted ; the other hemisphere is dark. As the Moon moves in its orbit round the Earth, the illuminated side is turned towards us in varying proportions; and the relation between the phases thus produced and the relative positions of Sun, Earth, and Moon is shown in the upper part of the plate.

When the Moon is but a few days old, and appears as a thin crescent, it frequently happens that part of the Moon which is not lit directly by the Sun is seen faintly shining by the reflected " Earth-light," an appearance known as " the old Moon in the new Moon's arms."

Eclipses of the Moon.—The Sun throws behind the Earth a dark cone of shadow, which reaches a long way beyond the path of the Moon, and if it had happened that the path of the Moon lay precisely in the ecliptic, then, at every full Moon, she would pass through this shadow, and be totally eclipsed. Since, however, the Moon's path makes a small angle with the ecliptic, she usually passes a little above or below the cone of shadow, and escapes eclipse. But, from time to time, the Moon is crossing the ecliptic just about the time of full, and then a partial or total eclipse occurs, and is visible over the whole of that hemisphere of the Earth which is at the moment turned away from the Sun and towards the Moon. It follows that a total eclipse of the Moon—being visible, whenever it occurs, over at least half the Earth—is not a very uncommon spectacle.

Eclipses of the Sun.—The Sun also throws behind the Moon a dark cone of shadow ; smaller than that thrown behind the Earth, because the Moon is smaller ; but just long enough, on the average, to reach the Earth. When the Moon is nearest the Earth, the cone of shadow may cover a space about 170 miles broad ; and, with the motion of the Moon, this shanow-patch sweeps quickly over the Earth. Within the shadow-belt, for a few minutes, the Sun is just a little more than completely obscured by the Moon, and there is a *total eclipse* of the Sun. When the Moon is farthest from the Earth, the shadow-cone does not reach the Earth, so that from no point can the Sun be seen completely obscured ; at best, there is a ring of Sun showing all round the Moon, and the eclipse is *annular*, At points lying outside the belt of totality, or of annularity, the Sun may be partially obscured by the Moon, and there is a *partial* eclipse. But the limits within which any eclipse at all is visible are far within the boundary of the whole hemisphere of the Earth which is turned towards the Sun, and consequently eclipses of the Sun of any kind are much more rarely seen at any one place than are eclipses of the Moon.

PLATE 15.

PATHS OF TOTAL ECLIPSES OF THE SUN, 1901—1950.

In his great work, "Canon der Finsternisse," Prof. Oppolzer has given maps of the paths of the Moon's shadow over the surface of the Earth for all the total and annular eclipses of the Sun between the years 1207 B.C. and 2162 A.D. From this work Plate 15 has been prepared, showing the tracks of the *total* eclipses visible between 1901 and 1950 A.D. At the western end of each line the eclipse begins at sunrise ; the point in the middle of each line, where the eclipse is at noon, is marked by a circle ; and at the eastern end of the line the eclipse begins at sunset.

An examination of these curves will show in a striking way the repetition of eclipses in a period of about eighteen years and eleven days, which period, known to the Chaldeans, is called the *Saros*. Take, as an example, the Great Eclipse of 1901—May 18th, occuring at mid-day in long. 97° E., lat. 2° 5'. We have on our Plate three eclipses of this series, viz. :—

			Lat.	Long.
1901—May 18	...	Eclipse at mid-day in 97° E.	2° S.	
1919— „ 29	...	„ „ 18 W.	4 N.	
1937—June 8	...	„ „ 131 W.	10 N.	

And later eclipses of the same series are :—

1955—June 20	...	Eclipse at mid-day in 117° E.	15° N.	
1973— „ 30	...	„ „ 6 W.	19 N.	
1991—July 11	...	„ „ 105 W.	22 N.	

The centre of the track of the eclipse is gradually moving north, and is at each repetition about seven hours, or 105°, farther west in longitude.

Let us take as another example the history of the eclipse which will be total in England in June, 1927. The dates of the three eclipses of this series represented on our Plate are :—

1909—June 17. | 1927—June 29. | 1945—July 9.

The first begins in Siberia, crosses close to the North Pole, and runs down the west coast of Greenland. The second begins in the Atlantic, south-west of Ireland, crosses Great Britain, runs up Norway, through the Arctic Ocean, and ends south of Behring Straits. The third begins in Canada, crosses Greenland and northern Norway, and ends in Central Asia.

It will be seen that the circumstances of the path of an eclipse are very complex, especially when its centre is in high latitudes, and the reason for this may be readily understood if one looks at a globe, tilted with respect to the Sun, according to the time of the year of the eclipse, and considers how the shadow of a body, the Moon, passing between the Sun and the globe would cut across the tilted lines of latitude and longitude.

PLATE 16.

TYPICAL SOLAR CORONAE.

By far the most beautiful feature of the totally-eclipsed Sun is the *corona* of pale white light which flashes out as soon as the dazzling photosphere is completely covered by the

Moon. In the three or four minutes which is the average duration of totality it is almost impossible to draw or describe the very complex structure of this appendage of the Sun. But, for the last twenty-five or thirty years, almost every eclipse has been successfully photographed. A reference to Plate 15 will show the arduous character of the journeys which are often involved in eclipse-observation. The six photographs which are reproduced in Plate 16 were taken as follows :

1.	1871.	Dec. 12.	H. Davis.	Baikul, India.
2.	1882.	May 17.	Abney aud Schuster.	Egypt.
3.	1893.	April 16.	J. Kearney.	Fundium, W. Africa.
4.	1878.	July 29.	W. Harkness.	Wyoming.
5.	1889.	Jan. 1.	W. H. Pickering.	California.
6.	1900.	May 28.	E. E. Barnard.	Wadesborough, N. Carolina.

The photographs have been arranged in two sets, in which it will be seen that the corona is of distinctly different types. In the first set—1871, 1882, 1893—the corona is fairly equally distributed right round the limb of the Sun ; in the second set—1878, 1889, 1900—the corona has large equatorial extensions, and at the poles it is broken up into short, distinct streamers. Further, it will be noticed that the interval between successive photographs in each set is about eleven years—the sun-spot period ; the first set fall near the times of sun-spot maximum, the second near times of sun-spot minimum.

Not very much is known of the nature of the corona. The streamers shine largely, if not entirely, by reflected light from the Sun, and must, therefore, be composed of small particles. Diffused amongst them, but probably not sharing in their radial structure, is an unknown gas—called, for convenience, "coronium." That some of the detailed structure of the corona is connected with underlying sun-spots and prominences is certain. But the most significant fact is the evident dependence of the forces which determine the form of the corona upon the same cause, whatever it may be, which produces the periodicity of the sun-spots, disturbance of the magnet, and aurorae.

The corona and prominences alike are ordinarily invisible to us, because they are not nearly so bright as the flare in our atmosphere which seems to surround the Sun. The spectroscope has made it possible to observe the prominences continuously ; but, up to the present, no method has been found of viewing the corona except during the rare minutes of a total eclipse.

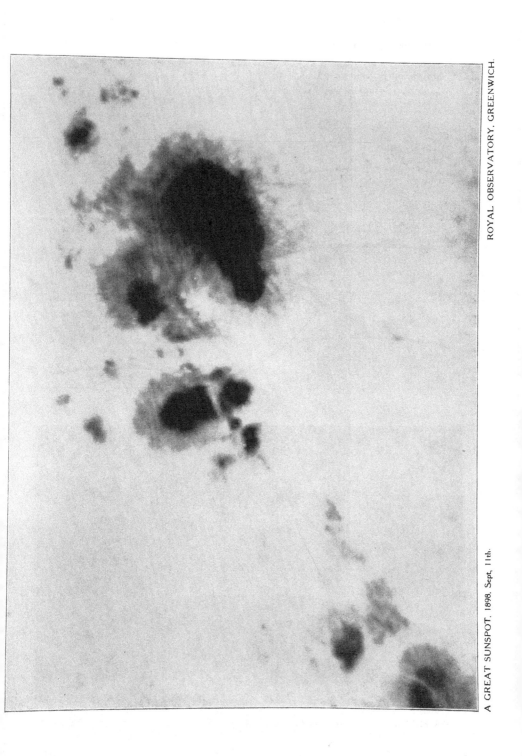

A GREAT SUNSPOT. 1898. Sept. 11th.

ROYAL OBSERVATORY. GREENWICH.

Plate 12.

SOLAR PROMINENCES. Drawn with the
spectroscope by J. FENYI, 1895, July 15th.

SOLAR PROMINENCES. Photographed by E. E. BARNARD during the Total Eclipse of the Sun, 1900, May 28th.

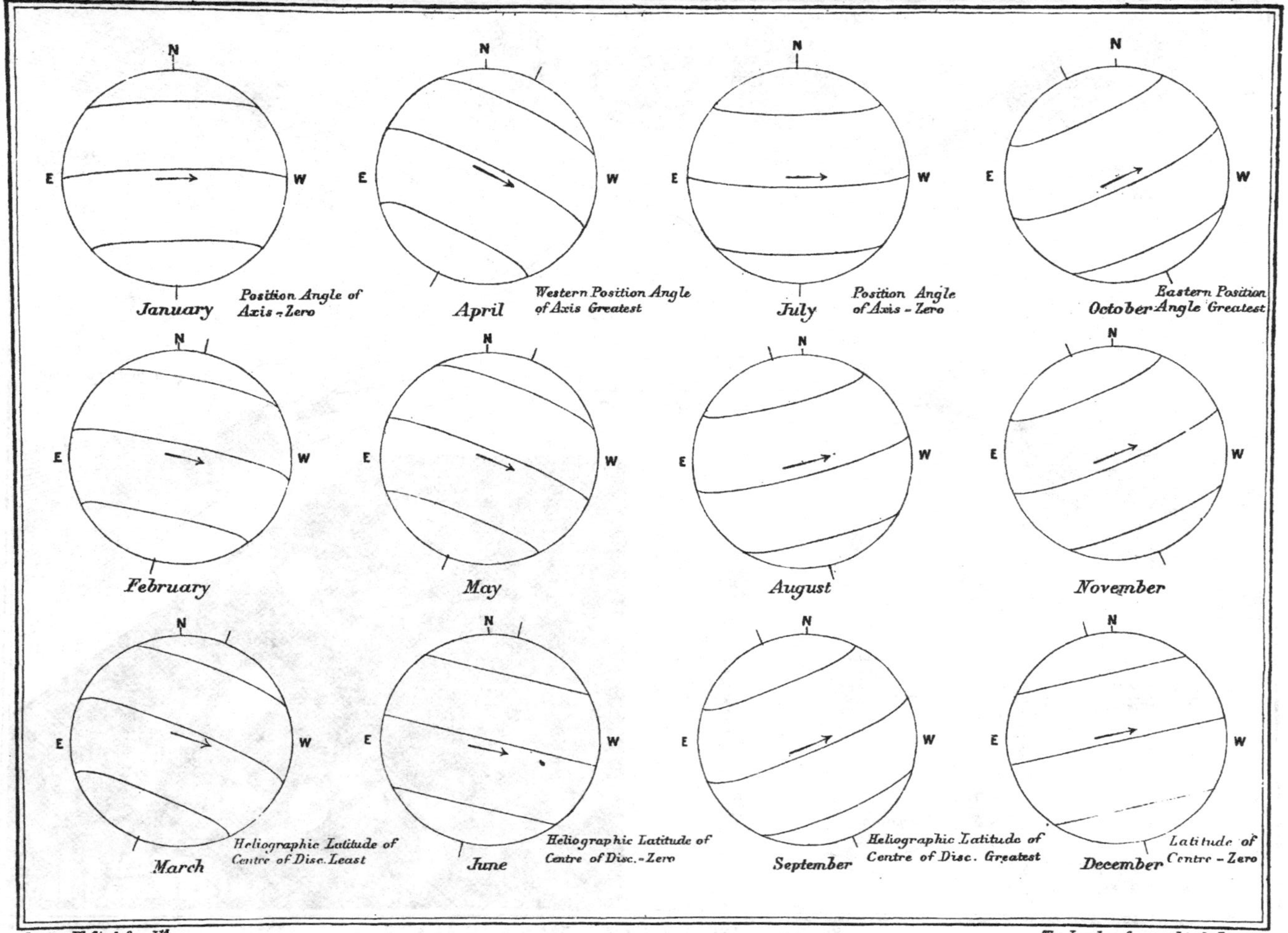

Plate 13

January
Position Angle of
Axis = Zero

April
Western Position Angle
of Axis Greatest

July
Position Angle
of Axis = Zero

October
Eastern Position
Angle Greatest

February

May

August

November

March
Heliographic Latitude of
Centre of Disc. Least

June
Heliographic Latitude of
Centre of Disc. = Zero

September
Heliographic Latitude of
Centre of Disc. Greatest

December
Latitude of
Centre = Zero

George Philip & Son, L^td.

The London Geographical Institute.

Plate 14

ECLIPSES AND PHASES OF THE MOON

CORRESPONDING PHASES

Relative Positions of Sun, Earth, & Moon.

Direction of the Sun

New Moon

First Quarter

Earth

Last Quarter

Full Moon

ECLIPSES

SUN

Annular Eclipse of Sun

Total Eclipse of Sun Minimum Totality

Moon

Moon

Ecliptic

Earth

Shadow

Moon

Partial Eclipse

Earth's
Shadow

Path of border

Total Eclipse

Earth's
Shadow

No Eclipse

Total Eclipse of Sun

Moon

Eclipse of Moon.

Earth

The London Geographical Institute

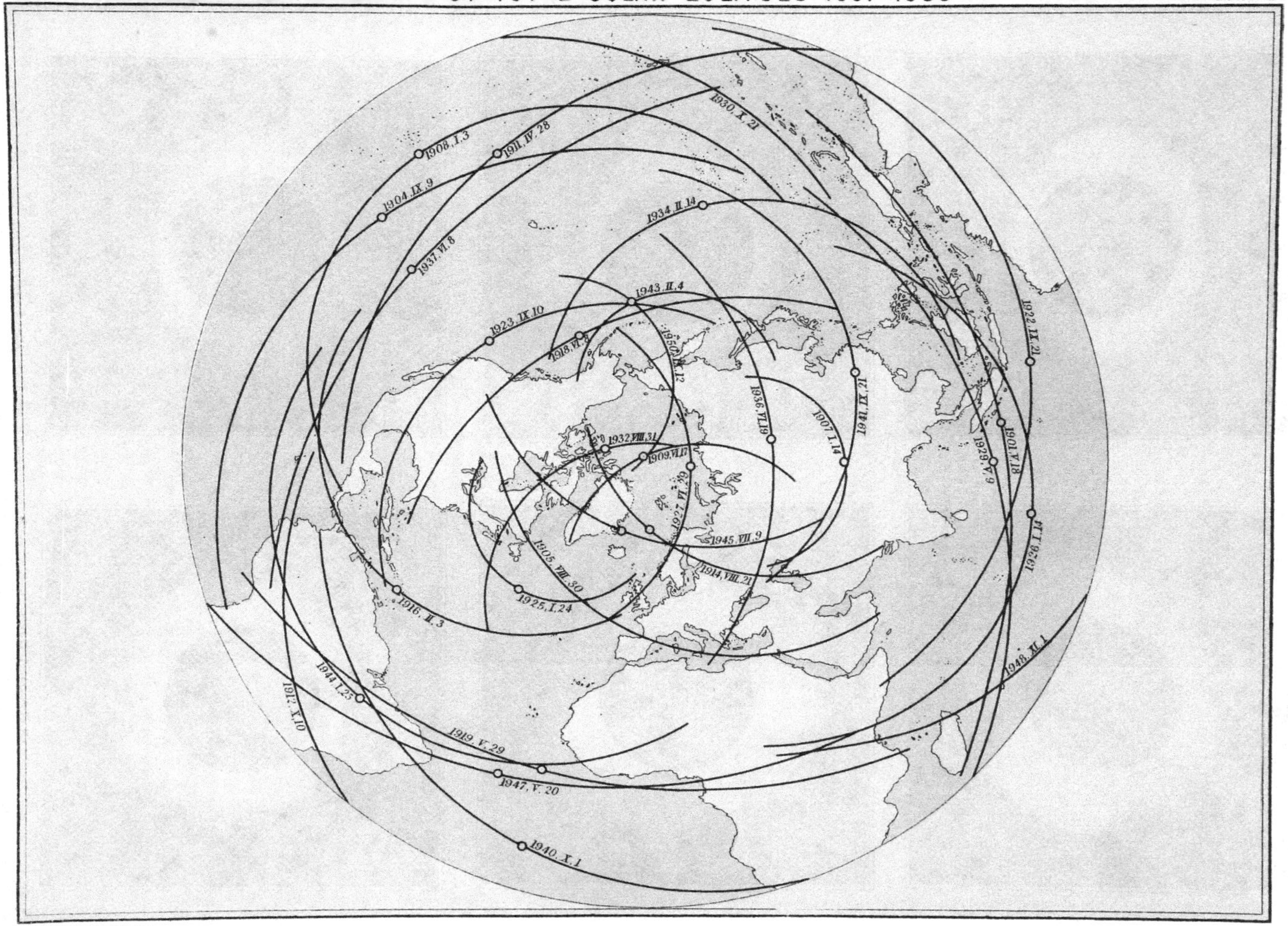

George Philip & Son, L^td.

The London Geographical Institute

Plate 16.

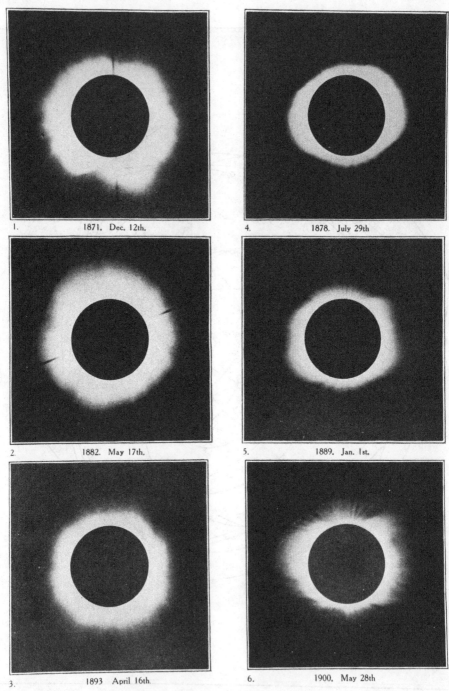

1.   1871. Dec. 12th.

4.   1878. July 29th

2.   1882. May 17th.

5.   1889. Jan. 1st.

3.   1893 April 16th.

6.   1900. May 28th

THE SOLAR CORONA. Photographed during total eclipses of the Sun.

CHAPTER IV.

PLATE 17.

DONATI'S COMET.

This, the most famous comet of the 19th Century, was discovered by Donati at Florence, on June 2nd, 1858, as a small telescopic object approaching the Sun. Not for nearly three months did it become visible to the naked eye, but thence, right up to the time of its perihelion passage, at the end of September, it grew rapidly in brightness until its starlike nucleus was as bright as the Pole star. During September its tail was directed nearly towards the Earth, and, though bright, was seen so much foreshortened that its effect was greatly marred; but as the comet passed perihelion and began to recede from the Sun, its path, by good fortune, was most favourably placed. The splendid plumed tail then lay almost at right angles to the line of sight, and its whole length was for the first time displayed. Other comets have had longer tails, though this was more than forty million miles long, but none have surpassed Donati's comet in beauty. The main tail, the curved plume, was of the type shown afterwards by the spectroscope to consist of hydrocarbons; the thin straight streamers are of the hydrogen type. Evaporated, apparently, from the nucleus of the comet by the heat of the Sun, the particles of the tail are repelled from the Sun by some force whose nature is still problematical, and driven backwards from it with a speed which must be comparable with that of the speed of light itself.

On the evening of October 5th, Donati's comet was at its best, when its motion involved the bright star Arcturus in the brightest part of its tail, through which the star shone undimmed. Our plate, which was drawn by Prof. Bond, at the Harvard College Observatory, shows Arcturus close to the comet's head, while its tail sweeps up between the Great Bear and the Northern Crown.

PLATE 18.

No. 1.

HOLMES' COMET AND THE ANDROMEDA NEBULA.

On Nov. 6, 1892, Mr. Edwin Holmes discovered in London a comet which was in many ways remarkable. When found it was close to the great nebula in Andromeda, and its motion was so slow that, throughout the month of November, it could be photographed on the same plate with the nebula. Plate 18 is a reproduction of a photograph taken at the Lick Observatory, on November 10th, by Professor Barnard, who describes the comet as "round, and sharply defined like a planetary nebula, with a symmetrical, nebulous atmosphere surrounding it for some distance."

The after-history of this comet is very curious. By the middle of December, it had grown so exceedingly faint and ill-defined that scarcely any telescope could show it. But, in the middle of January, it suddenly brightened up, and condensed into a small, hazy, star-like object, after which it again became diffuse, and finally vanished.

The comet's orbit was equally remarkable. It lay entirely between Mars and Jupiter, in the zone of the minor planets ; and it has even been suggested that the comet was not a comet at all, but the result of some celestial accident—such as a collision—which had befallen an asteroid.

Nos. 2 AND 3.

COMET a 1893, IV. (BROOKS.)

This comet, though small—and, as a visual object, insignificant—was, in some ways, the most remarkable comet that has yet been studied by photography. The plate is a reproduction of part of a series of photographs taken by Professor Barnard at the Lick Observatory. The motion of the comet was towards the north-east, the left-hand top corner of the picture. On 1893, Oct. 20th, the tail was straight, but gradually widening towards the end ; on the next day, the date of the second picture, it had been completely transformed. The tail is very much distorted, as if the matter of which it is formed had encountered some resistance. On the following day, October 22nd, the tail was completely wrecked, and large portions of it were detached. In our ignorance of the way in which a comet's tail is produced and maintained, it is scarcely possible to say anything definite by way of explanation of these changes. That the comet had encountered some resisting medium is a plausible conjecture, but nothing more.

PLATE 19.

COMET 1901. I.

The Great Comet of 1901, visible in the Southern Hemisphere, was by far the finest comet that had been seen for twenty years. It appeared very suddenly on April 24th, and was discovered independently by several persons in South Africa and Australia. It was then at perihelion, and visible only just before sunrise, but during the succeeding days it passed, apparently, still closer to the Sun, and was lost in the daylight. By May 3rd it was sufficiently clear of the Sun to be visible in the evening twilight, and on May 4th the photograph, from which Plate 19 is made, was taken at the Royal Observatory, Cape of Good Hope, with the Victoria telescope, in twilight. The tail is noticeably unsymmetrical, streaming from each side of the nucleus, but much more strongly on the south-west side. About this time there appeared on the same side a long, straight, faint tail, making an angle of about 30° with the axis of the main tail, and as the comet got away from the Sun into darker sky, this tail could be traced for about 25°, the extreme length of the main tail being about 7°.

COMET b 1902. III. (PERRINE.)

This was an excellent example of the kind of comet which raises false hopes when it is reported in the papers as "visible to the naked eye." At its brightest it was little more conspicuous than the Andromeda nebula, with which few people are familiar as a naked-eye object ; in the telescope, it was an almost formless patch of light, with a vague tail. The photograph—taken at the Royal Observatory, Greenwich, on Sept. 29, 1902—shows the tail strongly cleft. Six divisions can be counted in the original from which the plate was made.

This photograph was made with an exposure of 62m. The comet was in rapid motion amongst the stars, and the telescope with which the photograph was made was kept pointed precisely to it ; in consequence of this, the stars appear as trails, and give a precise idea of the amount by which the comet had moved during the hour which was needed to secure this picture.

George Philip & Son, Ltd.

The London Geographical Institute

HOLMES' COMET AND THE ANDROMEDA NEBULA.

1892. Nov. 21st.

1893. Oct. 21st.

BROOKS' COMET. 1893.IV.

From Photographs by E. E. BARNARD.

6-in. Portrait Lens, LICK OBSERVATORY.

1893. Oct. 21st.

Plate 18.

PERRINE'S COMET. 1902. III

30-in. Reflector, ROYAL OBSERVATORY, GREENWICH.

GREAT COMET OF 1901

May 4th   McClean Telescope, ROYAL OBSERVATORY, CAPE OF GOOD HOPE.

CHAPTER V.—THE MOON.

PLATES 20, 21 & 22.

The three photographs here reproduced were taken at the Yerkes Observatory with the great telescope, temporarily converted into a photographic telescope by the device of photographing through a screen of yellow glass in contact with the plate.

Plate 20 shows the region of the Mare Serenitatis and the Apennines. The Mare is more than 400 miles across, and is singularly free from Craters. The appearance in the photograph of the curious serpentine ridge towards its western border is a good example of the importance of selecting the right moment for studying any particular lunar object. When the photograph was taken this ridge was conspicuous : had it been taken a few hours later the ridge would have disappeared. It is really very low, so that it soon loses its shadow, and as soon as that happens it is no longer distinguishable.

The bright, white spot Linné has a long history. It was drawn by old observers as a deep crater. For many years it has been merely a bright spot, with scarcely any depression at all. Opinions differ widely as to the reality of any change ; perhaps, on the whole, the evidence is in favour of something having happened. But the doubt as to the trustworthiness of the old observations emphasises the value of photographs such as these, which could scarcely give a wrong verdict on such a point.

Craters differ much in their brightness ; Alfraganus and Dionysius have exceptionally brilliant walls ; Julius Cæsar and Boscovich are very dark.

The boundary of Julius Cæsar towards Sosigenes has a broken down and denuded appearance ; the deep valley alongside it has probably been formed by the fusion of several craters, which are frequently found three or four in a row close together.

The Apennines and the Caucasus of the Moon are mountainous regions much more resembling those of the Earth than do the lunar mountains in general. The peaks run up to 18,000 and 20,000 feet, and the N.E. boundary of the Apennines is a very steep cliff, not well shown in the photograph, which shows it under a setting sun.

There is a curious contrast between the craters Archimedes and Aristillus. The former, though 50 miles in diameter, has its crater floor only some 600 feet below the plain outside. Its walls, about 5,000 feet high, look broken and denuded, and the crater has the appearance of having been filled up nearly to the brim by an outflow of lava. In Aristillus, on the contrary, the depth from brim to floor is 11,000 feet ; the central peak and terraces are preserved, and the plain all round is covered as if with the debris of relatively late eruptions.

In Plate 21 we have a picture of the most rugged and broken part of the Moon's surface. The crater Tycho at sunrise, as shown here, is relatively undistinguished, though of such size that Mont Blanc would stand on its floor, and from its summit it would not be possible to see over the crater wall. But as the Moon gets toward full, while most of the other craters become hard to see—Clavius, for example, almost entirely disappears under the perpendicular illumination—Tycho stands out conspicuously brilliant, the centre of a system of radiating bright streaks, whose nature is a mystery. They go straight across mountains and plains ; there is only one well-marked case, Saussure, in which the streak seems to turn aside to avoid a mountain.

It is curious that in the district around Clavius the western walls of the craters are generally higher than the eastern. In Clavius itself, a peak of the western wall stands 17,000 feet above the floor, and the deepest of the smaller craters within is 6,000 feet deep.

At the extreme bottom of the picture, below and to the left of Pitatus, is the Straight Wall, recognised only by its shadow. The wall is almost perfectly straight, 60 miles long, and about 1,000 feet high. It is much steeper on the east than on the west side, and is, perhaps, better called a cliff than a wall.

PLATE 22.—In Copernicus and the region around it we find lunar scenery on the grandest scale. The crater itself is about 60 miles in diameter ; the highest peak is more than 12,000 feet above the floor ; the central mountain above 2,000 feet high. The successive terraces of the wall are said to resemble those of the crater of Teneriffe : the ridges running down on to the plain suggest outpourings of lava. To the north is Mt. Carpathus with an enormous c eft. To the west the whole plain is riddled like a sieve with small craters. The line of these small craters running north and south, and becoming at the north end a deep cleft, suggests the question : Are these small craters formed along a pre-existing cleft, or is the cleft, as we see it, formed by the amalgamation of a number of small craters in a line ?

PLACE OF THE MOON.

From the monthly maps, 39—50, the positions of the Moon at different periods in the lunation can be learned. In the first place, it is to be noted that our Satellite lies always in or close to that part of the sky marked as the " Track of the Planets." When it is full the Moon is in opposition, and comes on the meridian at midnight, and hence we have the following rule :

Look out the monthly map for the month in question, then the full Moon lies in that part of the heavens where the " Track of the Planets " crosses the central meridian, already defined to be the line drawn on the map from the North point to the South point.

Example 1.—In what Constellation does the full Moon appear in September ?

Solution.—The answer is given by Plate 47, where the "Track of the Planets " crosses the central meridian in Pisces, which indicates the required position.

Example 2.—When is the full Moon near the Pleiades ?

Solution.—Plate 49 shows the Pleiades on the central meridian, and accordingly November is the answer to the question.

To find the position of the Moon at the time of the first quarter, the following is the method.

Look out the monthly map for three months *preceding* the given date, then the constellation in or near which the Moon lies at the first quarter is shown at the intersection of the "Track of the Planets " with the central meridian.

Example.—In what constellation does the first quarter Moon appear in June ?

Solution.—The map three months earlier is Plate 41 for March. This shows the intersection of the " Track of the Planets " and the central meridian in Virgo, which is accordingly the answer required.

To find the position of the Moon at the time of the last quarter, the following is the method.

Look out the monthly map for three months *following* the given date, then the Constellation in or near which the Moon lies at the last quarter is shown at the intersection of the " Track of the Planets " with the central meridian.

Example.—In what constellation does the last quarter Moon appear in July ?

Solution.—The map three months later is Plate 48, which shows that the constellation is Aries.

It ought to be observed that, on account of the rapid motion of the Moon, only a rough indication of its place can be expected from the process here given, and that the accuracy will be greater the nearer the phase in question happens to the middle of the month.

The foregoing problems can also be solved by the more general method now to be described. The Table of Moon Age shows the position in the heavens which the Moon occupies at any age in any month. The use of this Table is as follows.

Enter the table in the verticle column bearing the name of the month. Then take the age in that column nearest the given age, and the figure at the left on the same row gives the number of the monthly map in which the region where the Moon is situated lies on the "central meridian" where the "Track of the Planets" crosses it.

THE TABLE OF MOON AGE.

Map.	Jan.	Feb.	March.	April.	May.	June.	July.	Aug.	Sept.	Oct.	Nov.	Dec.
39	14	12	10	7	5	3	29	25	23	20	18	16
40	17	14	12	10	7	5	3	27	25	23	21	19
41	19	16	15	12	10	8	5	2	28	25	23	21
42	22	19	17	14	12	10	8	5	1	0	26	24
43	25	22	21	16	14	12	10	7	4	2	28	26
44	27	25	23	18	16	14	11	9	7	4	2	29
45	29	27	25	20	18	16	14	11	9	7	5	2
46	2	0	27	25	21	18	16	14	11	9	7	4
47	5	2	29	27	24	20	18	16	14	11	10	7
48	7	4	3	0	27	23	20	18	16	14	12	9
49	10	7	5	3	0	27	23	20	18	16	14	11
50	12	9	8	5	2	0	27	23	20	18	16	14

Example 1.—Where does the Moon lie when four days old in October ?

Solution.—The October column in the Table of Moon Age being referred to, the sixth figure from the top gives 4, the age of the Moon, and the figure at the end of that row on the left is 44. This monthly map shows that the Moon must then be in or near Sagittarius.

Example 2.—What will be the age of the Moon when on the meridian at 10 P.M. in August ?

Solution.—At 10 P.M. in August, the heavens will be as in Plate 45. Therefore we refer to the row for Map 45 in the Table of Moon Age, which shows, under the column August, that the moon must then be about 11 days old.

Example 3.—Determine when the Moon, at the first quarter, has a specially high altitude.

Solution.—The heavens must be as in Plate 49, which refers us to the last row but one of the Table. For the Moon to be 7 days old we look under the column February, in which month the heavens are as in Plate 49 about 6 P.M.

PLATES 23 TO 38.

THE LUNAR OBJECTS.

For the study of the Lunar formations, Plates 23 to 38 have been specially drawn.

As the astronomical telescope shows the Moon turned upside down, and with right and left interchanged, the maps of our Satellite are represented accordingly. The four quadrants (Plates 23, 24, 25, 26) are designated in the manner shown in the annexed figure. For observations of the Moon, the "terminator" or boundary between light and shade, is the place where the objects are best seen, and Plates 23—38 of the present Atlas have been arranged to facilitate observation of the Lunar formations on the terminator at various ages, from new to full. The terminators for each day of a lunation are marked on the quadrants; the morning terminator being that when the Sun is rising on the object in question. The quadrants also enable the latitudes and longitudes of Lunar objects to be found.

As the Moon is so much more conveniently observed from new to full, than from full to new, it is the former series of changes that have been more particularly provided for. The telescopic view of the Crescent Moon, 3 days old, is shown in Plate 27. On the opposite page an index outline is given on

Moon in Inverting Telescope.

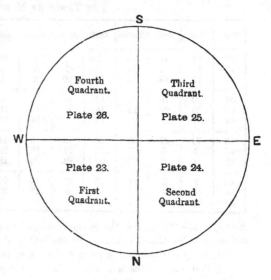

which each of the formations receives a special number or letter. The name of the formation may be found by looking out the number or letter in the Catalogue of Lunar formations; but for greater convenience in reference, the names of the chief objects visible in each phase are set out on the Index outline as well. As the Moon grows day by day, the terminator changes, and an ever varying series of objects is presented. A special Plate is therefore given for each day of the Moon's age, from the 3rd up to the 14th, when the Moon is full. Before the third day the Moon is so close to the Sun that observations cannot be made with advantage.

Suppose, for instance, that the Moon is 9 days old. The observer then refers to Plate 33. On the terminator, a little below the middle, he notes a fine crater, and desires to learn its name. The Index outline assigns the Number 380, and the list on the margin shows that this feature is named "Copernicus." The observer will be able to trace the same object with lessening detail up to the time of Full Moon. See Plates 34 to 38. From the comparison of any one of these Plates with the figure on this page, it appears that Copernicus must lie in the "Second Quadrant" or on Plate 24, where the great crater will be found again as No. 380, a conspicuous object at 20° East longitude, and 10° North latitude. Along the top of Plate 24 are shown the positions of the terminators at corresponding ages of the Moon. It will be noted that the morning terminator on the 9th day passes through Copernicus. So also does the evening terminator on the 24th, so that if the observer desires to study Copernicus when illuminated by the sunlight from the opposite side, he may repeat his observation 15 days later.

As another illustration, let us suppose the Moon to be 4 days old, and that after comparing the Moon with Plate 28 we desire to know the name of that large round dark patch, a little below the centre, which lies midway between the limb and the terminator. The Index outline shews it marked A, and from the reference to the margin or to the Catalogue the object is identified as the Mare Crisium. It is represented in Plate 23 as A. near the top at the left.

To show the mode of representing the ranges of Lunar mountains, we may suppose the student to be looking at the Moon a little after the first quarter, say on the eighth day, as on Plate 32. He notices a remarkable formation a little below the centre. The Index outline labels this object c, and the margin shows that we are looking at the lunar Apennines. Plate 24 exhibits the Apennines pointing towards Copernicus.

Suppose that a view of some particular formation of known name be specially desired, the process is as follows. Look it out in the Index at the end of this volume, the first reference is to the quadrant, and the next is to the plate where the object is represented on the terminator.

Thus, for instance, to find the position of Plato. The Index shows first of all that it lies on Plate 24, that is, in the Second Quadrant. The next reference is to Plate 32, which shows the object lying near the terminator when the Moon is 8 days old. There are further references to 33, 34, and 35, where the object is also visible. The evening terminator on Plate 24 shows that when this object is suitably placed for observations with the opposite illumination, the Moon is about 23 days old. The subsequent references in the Index are to those pages of the Introduction in which the object is mentioned.

The beginner should, however, be apprised that even with the assistance which it is hoped that these maps will afford him, considerable pains are often required to identify the lunar objects. In the first place, various causes produce what are known as librations of the Moon, whose effect is that the Moon does not always turn precisely the same face toward us. The maps are accommodated to a state of mean libration, and the student must not be surprised if he finds an object sometimes higher and sometimes lower than its position in the map would have led him to expect. These changes often produce considerable variations in the appearance of the lunar formations. It must also be remembered that the age of the Moon cannot be always exactly that of the map which comes nearest to it. This will often involve considerable alterations in the appearance of the lunar formations from those which they present at the exact phase which the map depicts. The elucidation of the several points which thus arise will afford much interesting occupation, and will, it is hoped, lead the student to a close acquaintance with the beautiful scenery of our Satellite.

CATALOGUE OF LUNAR OBJECTS.

Figures refer to the Number of the Crater or similar formation, capital letters refer to the so-called "Seas," and small letters refer to the Mountain Ranges and isolated Mountains.

1 Langrenus.	47 Steinheil.	93 Pons.
2 Kästner.	48 Vlacq.	94 Pontanus.
3 Vendelinus.	49 Rosenberger.	95 Gemma Frisius.
4 Maclaurin.	50 Nearchus.	96 Poisson.
5 Hecatæus.	51 Hommel.	97 Aliacensis.
6 Ansgarius.	52 Pitiscus.	98 Werner.
7 Petavius.	53 Mutus.	99 Apianus.
8 Wrottesley.	54 Manzinus.	100 Playfair.
9 Palitzsch.	55 Censorinus.	101 Blanchinus.
10 Hase.	56 Torricelli.	102 La Caille.
11 Legendre.	57 Capella.	103 Delaunay.
12 Wilhelm Humboldt.	58 Isidorus.	104 Faye.
13 Phillips.	59 Mädler.	105 Donati.
14 Furnerius.	60 Bohnenberger.	106 Airy.
15 Stevinus.	61 Rosse.	107 Argelander.
16 Snellius.	62 Fracastorius.	108 Parrot.
17 Adams.	63 Piccolomini.	109 Albategnius.
18 Marinus.	64 Stiborius.	110 Hipparchus.
19 Fraunhofer.	65 Riccius.	111 Halley.
20 Oken.	66 Rabbi Levi.	112 Hind.
21 Vega.	67 Zagut.	113 Horrocks.
22 Pontécoulant.	68 Lindenau.	114 Rhæticus.
23 Biela.	69 Nicolai.	115 Reaumur.
24 Hagecius.	70 Büsching.	116 Walter.
25 Boussingault.	71 Buch.	117 Nonius.
26 Boguslawsky.	72 Hypatia.	118 Fernelius.
27 Schomberger.	73 Delambre.	119 Stöfler.
28 Webb.	74 Theon Senr.	120 Faraday.
29 Messier.	75 Theon Junr.	121 Maurolycus.
30 Lubbock.	76 Taylor.	122 Barocius.
31 Godenius.	77 Alfraganus.	123 Clairaut.
32 Guttemberg.	78 Kant.	124 Licetus.
33 Magelhaens.	79 Theophilus.	125 Cuvier.
34 Colombo.	80 Cyrillus.	126 Bacon.
35 Cook.	81 Catharina.	127 Jacobi.
36 Santbech.	82 Tacitus.	128 Lilius.
37 McClure.	83 Beaumont.	129 Zach.
38 Crozier.	84 Descartes.	130 Kinau.
39 Bellot.	85 Abulfeda.	131 Pentland.
40 Borda.	86 Almanon.	132 Curtius.
41 Reichenbach.	87 Geber.	133 Simpelius.
42 Rheita.	88 Abenezra.	134 Miller.
43 Neander.	89 Azophi.	135 Schubert.
44 Metius.	90 Sacrobosco.	136 Apollonius.
45 Fabricius.	91 Fermat.	137 Firmicus.
46 Janssen.	92 Polybius.	138 Azout.

CATALOGUE OF LUNAR OBJECTS—*continued*.

139	Neper.	189	De la Rue.	239	Conon.
140	Condorcet.	190	Strabo.	240	Manilius.
141	Behaim.	191	Thales.	241	Ukert.
142	La Peyrouse.	192	Gärtner.	242	Triesnecker.
143	Hanno.	193	Democritus.	243	Hyginus.
144	Le Gentil	194	Arnold.	244	Agrippa.
145	Tannerus.	195	Moigno.	245	Godin.
146	Huggins.	196	Peters.	246	Ritter.
147	Timoleon.	197	Meton.	247	Sabine.
148	Zeno.	198	Euctemon.	248	Dionysius.
149	Schwabe.	199	Challis.	249	Manners.
150	Hansen.	200	Main.	250	Arago.
151	Alhazen.	201	Giòja.	251	Ariadæus.
152	Picard.	202	Scoresby.	252	Silberschlag.
153	Pierce.	203	Barrow.	253	De Morgan.
154	Taruntius.	204	W. C. Bond.	254	Cayley.
155	Secchi.	205	Christian Mayer.	255	Whewell.
156	Proclus.	206	Archytas.	256	Calippus.
157	Maskelyne.	207	Aristoteles.	257	Theætetus.
158	Jansen.	208	Eudoxus.	258	Cassini.
159	Vitruvius.	209	Alexander.	259	Aristillus.
160	Maraldi.	210	Egede.	260	Autolycus.
161	Cauchy.	211	Great Alpine Valley.	261	Mösting.
162	Einmart.	212	Grove.	262	Lalande.
163	Oriani.	213	Mason.	263	Herschel.
164	Plutarch.	214	Plana.	264	Ptolemæus.
165	Seneca.	215	Burg.	265	Alphonsus.
166	Macrobius.	216	Baily.	266	Arzachel.
167	Cleomedes.	217	Daniell.	267	Alpetragius.
168	Tralles.	218	Posidonius.	268	Lassell.
169	Burckhardt.	219	Chacornac.	269	Davy.
170	Hahn.	220	Le Monnier.	270	Gueriké.
171	Berosus.	221	Roemer.	271	Parry.
172	Gauss.	222	Bond.	272	Bonpland.
173	Geminus.	223	Maury.	273	Fra Mauro.
174	Bernouilli.	224	Littrow.	274	Thebit.
175	Messala.	225	Newcomb.	275	Straight Wall.
176	Berzelius.	226	Dawes.	276	Birt.
177	Hooke.	227	Plinius.	277	Purbach.
178	Schumacher.	228	Ross.	278	Regiomontanus.
179	Struve.	229	Maclear.	279	Hell.
180	Mercurius.	230	Sosigenes.	280	Pitatus.
181	Franklin.	231	Julius Cæsar.	281	Hesiodus.
182	Cepheus.	232	Boscovich.	282	Gauricus.
183	Oersted.	233	Taquet.	283	Wurzelbauer.
184	Shuckburgh.	234	Menelaus.	284	Sasserides.
185	Chevallier.	235	Sulpicius Gallus.	285	Ball.
186	Atlas.	236	Bessel.	286	Lexell.
187	Hercules.	237	Linné.	287	Nasireddin.
188	Endymion.	238	Aratus.	288	Orontius.

CATALOGUE OF LUNAR OBJECTS—*continued.*

289 Pictet.
290 Saussure.
291 Tycho.
292 Heinsius.
293 Wilhelm I.
294 Longonfontanus.
295 Street.
296 Maginus.
297 Deluc.
298 Clavius.
299 Cysatus.
300 Moretus.
301 Short.
302 Newton.
303 Gruemberger.
304 Cabeus.
305 Casatus.
306 Klaproth.
307 Wilson.
308 Kircher.
309 Bettinus.
310 Zuchius.
311 Segner.
312 Blancanus.
313 Scheiner.
314 Weigel.
315 Rost.
316 Bailly.
317 Schiller.
318 Bayer.
319 Pingré.
320 Hausen.
321 Phocylides.
322 Wargentin.
323 Schickard.
324 Drebbel.
325 Inghirami.
326 Hainzel.
327 Lehmann.
328 Lacroix.
329 Piazzi.
330 Lagrange.
331 Fourier.
332 Vieta.
333 Doppelmayer.
334 Lee.
335 Vitello.
336 Clausius.
337 Capuanus.
338 Cichus.

339 Mercator.
340 Campanus.
341 Kies.
342 Bullialdus.
343 Lubiniezky.
344 Nicollet.
345 Hippalus.
346 Agatharchides.
347 Gassendi.
348 Herigonius.
349 Letronne.
350 Mersenius.
351 Cavendish.
352 Byrgius.
353 Eichstädt.
354 De Vico.
355 Ramsden.
356 Billy.
357 Hansteen.
358 Sirsalis.
359 Fontana.
360 Zupus.
361 Crüger.
362 Rocca.
363 Grimaldi.
364 Damoiseau.
365 Riccioli.
366 Lohrmann.
367 Hermann.
368 Flamsteed.
369 Wichmann.
370 Euclides.
371 Landsberg.
372 Gambart.
373 Sömmering.
374 Schröter.
375 Pallas.
376 Bode.
377 Reinhold.
378 Hortensius.
379 Milichius.
380 Copernicus.
381 Stadius.
382 Eratosthenes.
383 Gay Lussac.
384 Tobias Mayer.
385 Kunowsky.
386 Encke.
387 Kepler.
388 Bessarion.

389 Reiner.
390 Marius.
391 Hevel.
392 Cavalerius.
393 Olbers.
394 Cardanus.
395 Krafft.
396 Vasco de Gama.
397 Seleucus.
398 Marco Polo.
399 Archimedes.
400 Beer.
401 Timocharis.
402 Lambert.
403 Pytheas.
404 Euler.
405 Diophantus.
406 Delisle.
407 Caroline Herschel.
408 Carlini.
409 Leverrier.
410 Helicon.
411 Kirch.
412 Piazzi Smyth.
413 Plato.
414 Timæus.
415 Birmingham.
416 Epigenes.
417 Goldschmidt.
418 Anaxagoras.
419 Fontenelle.
420 Philolaus.
421 Anaximenes.
422 J. J. Cassini.
423 Condamine.
424 Maupertuis.
425 Bianchini.
426 Sharp.
427 Mairan.
428 Foucault.
429 Harpalus.
430 J. F. W. Herschel.
431 Anaximander.
432 Pythagoras.
433 South.
434 Babbage.
435 Œnopides.
436 Robinson.
437 Cleostratus.
438 Xenophanes.

CATALOGUE OF LUNAR OBJECTS—*continued.*

439 Repsold.	445 Briggs.	451 Gruithuisen.
440 Harding.	446 Otto Struve.	452 Brayley.
441 Gérard.	447 Aristarchus.	453 Galileo.
442 Lavoisier.	448 Herodotus.	454 Horrebow.
443 Ulugh Beigh.	449 Wollaston.	
444 Lichtenberg.	450 Schiaparelli.	

MOUNTAIN RANGES AND ISOLATED MOUNTAINS.

a	Alps.	*m*	Straight Range.
b	Caucasus.	*n*	Percy Mountains.
c	Apennines.	*o*	Harbinger Mountains.
d	Carpathians.	*p*	Hercynian Mountains.
e	Sinus Iridum Highlands.	*q*	Pico.
f	Hæmus.	*r*	Piton.
g	Pyrenees.	*s*	Mt. Argæus.
h	Altai Mountains.	*t*	Mt. Hadley.
i	Riphæan Mountains.	*u*	Laplace Promontory.
j	La Hire.	*v*	Mt. Huygens.
k	Mt. Taurus.	*w*	Mt. Bradley.
l	Teneriffe Range.		

Mountains near the Limb :—

D'Alembert Mts.—on the east limb, extending from S. lat. 19° to N. lat. 12°.

The Cordilleras—near the east limb, extending from S. lat. 23° to S. lat. 8°.

The Rook Mountains—on the east limb, extending from S. lat. 39° to S. lat. 16°.

The Doerfel Mountains—on the south-east limb, extending from S. lat. 80° to S. lat. 57°.

The Leibnitz Mountains extend from S. lat. 70° on the west limb to S. lat. 80° on the east limb.

Humboldt Mountains—on the west limb, extending from N. lat. 72° to N. lat. 53°.

MARIA or SEAS.

A	Mare Crisium.	N	Sinus Æstuum.
B	„ Fœcunditatis.	P	„ Medii.
C	„ Australe.	Q	Mare Nubium.
D	„ Humboldtianum.	R	Sinus Iridum.
E	„ Tranquillitatis.	S	Oceanus Procellarum.
F	„ Nectaris.	T	Mare Humorum.
G	Lacus Somniorum.	V	Palus Somnii.
H	„ Mortis.	W	Sinus Roris.
J	Mare Serenitatis.	X	Palus Nebularum.
K	„ Frigoris.	Y	Mare Smythii.
L	„ Imbrium.	Z	Palus Putredinis.
M	„ Vaporum.		

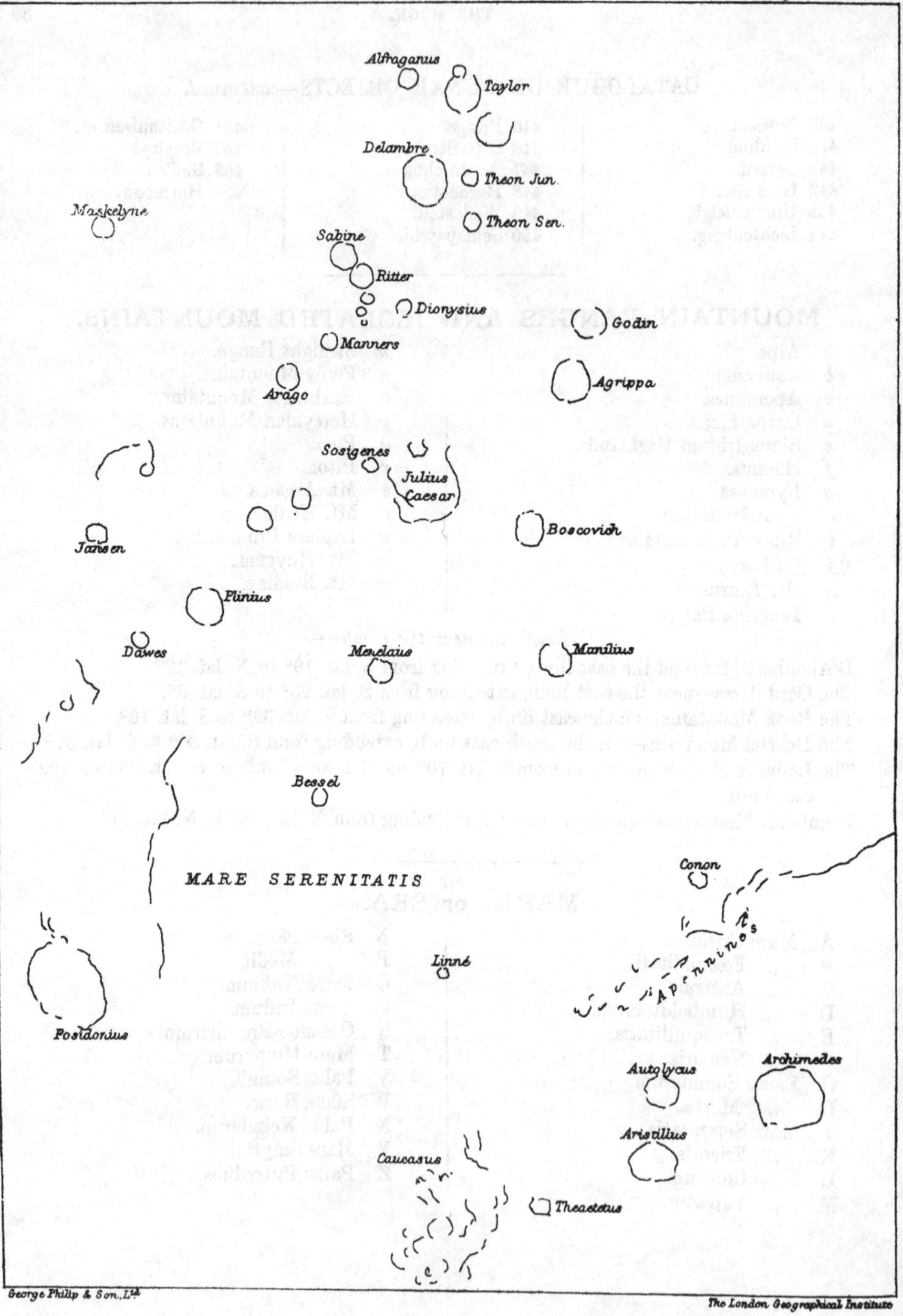

Alfraganus

Taylor

Delambre

Theon Jun.

Maskelyne

Theon Sen.

Sabine

Ritter

Dionysius

Godin

Manners

Agrippa

Arago

Sosigenes

Julius
Caesar

Boscovich

Jansen

Plinius

Dawes

Menelaus

Manilius

Bessel

Conon

MARE SERENITATIS

Apennines

Linné

Posidonius

Autolycus

Archimedes

Aristillus

Caucasus

Theaetetus

'HE MOON.   REGION OF THE MARE SERENITATIS. G. W. RITCHEY, 40-'n. Refractor, YERKES OBSERVATORY.

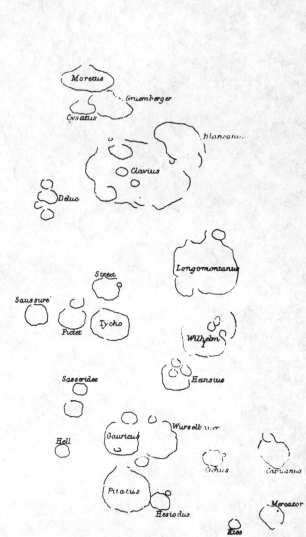

Plate 21.

THE MOON. REGION OF CLAVIUS AND TYCHO.   G. W. RITCHEY, 40-in. Refractor, YERKES OBSERVATORY.

Reinhold

Hortensius

Copernicus

Tobias Mayer

M. Carpathus

Gay Lussac

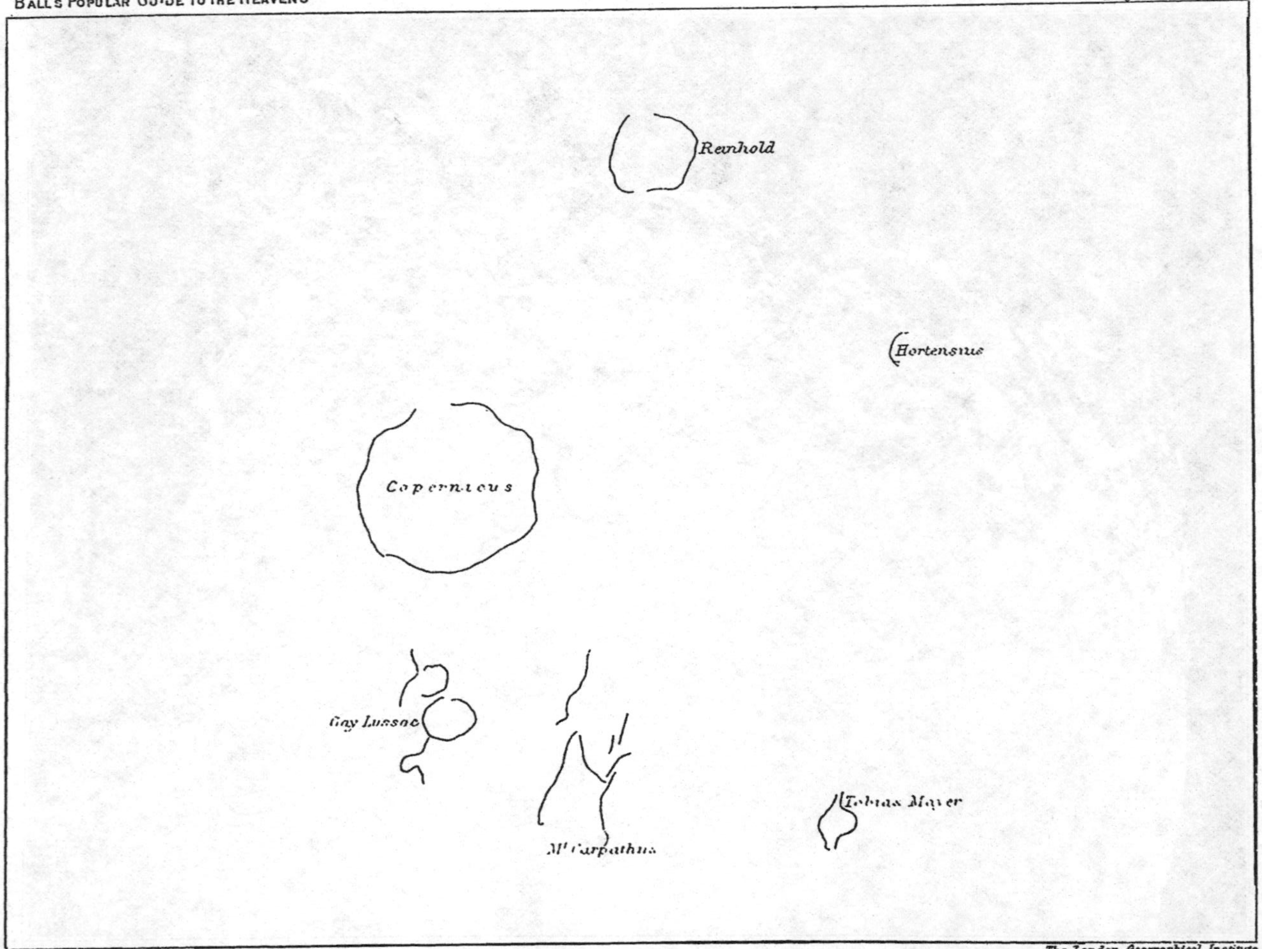

George Philip & Son, Ltd.

The London Geographical Institute

THE MOON. REGION OF COPERNICUS.

G. W. RITCHEY. 40-in. Refractor, YERKES OBSERVATORY.

Plate 23.

FIRST QUADRANT. N.W.

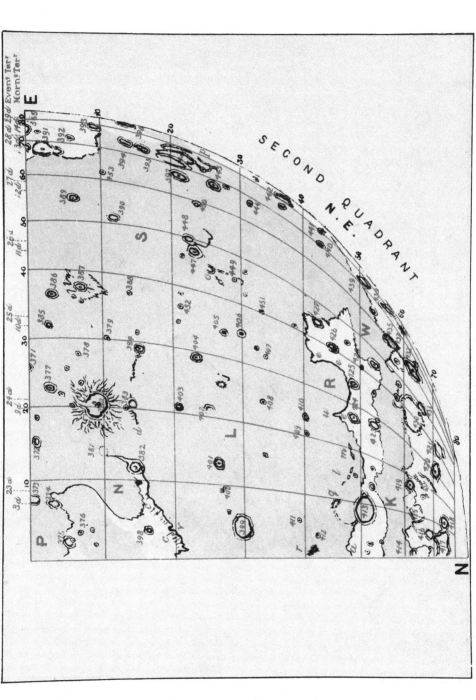

George Philip & Son, Ltd.

The London Geographical Institute.

Plate 25.

THIRD QUADRANT
S.E.

S

E

Morn⁹ Ter'r
Even'Ter'r

S

W

FOURTH QUADRANT S.W.

25. *Boussingault.*   14. *Furnerius.*   11. *Legendre.*

22. *Pontécoulant.*   12. *W. Humboldt.*   5. *Hecatæus.*

19. *Fraunhofer.*   7. *Petavius.*   3. *Vendelinus.*

                                                          1. *Langrenus.*

                                                          2. *Kästner.*

136. *Apollonius.*

137. *Firmicus.*

139. *Neper.*

140. *Condorcet.*

152. *Picard.*

151. *Alhazen.*

153. *Peirce.*

167. *Cleomedes.*

169. *Burckhardt.*

173. *Geminus.*

172. *Gauss.*

175. *Messala.*

180. *Mercurius.*

188. *Endymion.*

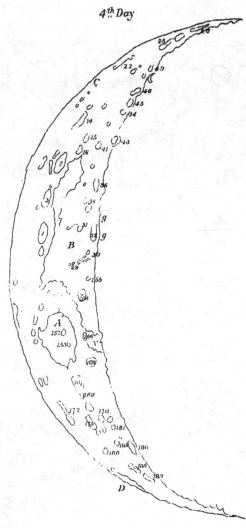

4ᵗʰ Day

A *Mare Crisium.*

B   ,,   *Fœcunditatis.*

C   ,,   *Australe.*

D   ,,   *Humboldtianum.*

26. *Boguslawsky.*
25. *Boussingault.*
22. *Pontécoulant.*

49. *Rosenberger.*
46. *Janssen.*
45. *Fabricius.*

44. *Metius.*
14. *Furnerius.*
15. *Stevinus.*
43. *Neander.*
41. *Reichenbach.*
16. *Snellius.*
7. *Petavius.*
36. *Santbech.*
34. *Colombo.*
3. *Vendelinus.*
31. *Godenius.*
32. *Guttemberg.*
1. *Langrenus.*
30. *Lubbock.*
29. *Messier.*
155. *Secchi.*
136. *Apollonius.*
154. *Taruntius.*
152. *Picard.*
153. *Pierce.*
156. *Proclus.*
166. *Macrobius.*
167. *Cleomedes.*
169. *Burckhardt.*
172. *Gauss.*
176. *Berzelius.*
175. *Messala.*
181. *Franklin.*
185. *Chevallier.*
186. *Atlas.*
180. *Mercurius.*
188. *Endymion.*
189. *De La Rue.*

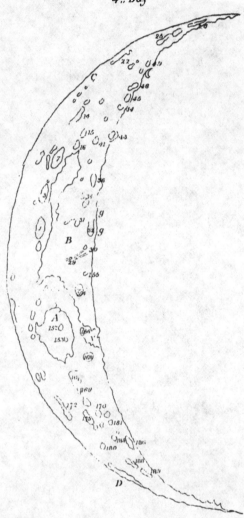

4ᵗʰ Day

A   *Mare Crisium.*
B   ,,   *Faecunditatis.*
C   ,,   *Australe.*
D   ,,   *Humboldtianum.*

V   *Palus Somnii.*

g.   *Pyrenees Mts.*

27. *Schomberger.*	49. *Rosenberger.*	46. *Janssen.*
54 *Manzinus.*	48. *Vlacq.*	45. *Fabricius.*
53. *Mutus.*	52. *Pitiscus.*	44. *Metius.*
		14. *Furnerius.*
		64. *Stiborius.*
		63. *Piccolomini.*
		62. *Fracastorius.*
		36. *Santbech.*
		7. *Petavius.*
		3. *Vendelinus.*
		31. *Godenius.*
		32. *Guttemberg.*
		1. *Langrenus.*
		58. *Isidorus.*
		57. *Capella.*
		29. *Messier.*
		159. *Vitruvius.*
		22. *Le Monnier.*
		221. *Roemer.*
		219. *Chacornac.*
		224. *Littrow.*
		218. *Posidonius.*
		214. *Plana.*
		187. *Hercules.*
		186. *Atlas.*
		188. *Endymion.*
		189. *De La Rue.*
		198. *Euctemon.*

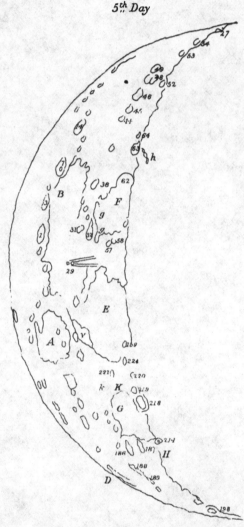

5ᵗʰ Day

A *Mare Crisium.*	D *Mare Humboldtianum.*	g. *The Pyrenees.*
B ,, *Fœcunditatis.*	G *Lacus Somniorum.*	k. *Taurus Mts.*
E ,, *Tranquillitatis.*	H ,, *Mortis.*	
F ,, *Nectaris.*		

126. *Bacon.*   71. *Buch.*   66. *Rabbi Levi.*
121. *Maurolycus.*   64. *Stiborius.*   67. *Zagut.*
                                 65. *Riccius.*   63. *Piccolomini.*
                                                               93. *Pons.*
               6th Day   62. *Fracastorius.*
                                                               90. *Sacrobosco.*
                                                               91. *Fermat.*
                                                               81. *Catharina.*
                                                               82. *Tacitus.*
                                                               80. *Cyrillus.*
                                                               59. *Mädler.*
                                                               79. *Theophilus.*
                                                               72. *Hypatia.*
                                                               247. *Sabine.*
                                                               246. *Ritter.*
                                                               250. *Arago.*
                                                               229. *Maclear.*
                                                               228. *Ross.*
                                                               227. *Plinius.*
                                                               226. *Dawes.*
                                                               236. *Bessel.*
                                                               220. *Le Monnier.*
                                                               219. *Chacornac.*
                                                               218. *Posidonius.*
                                                               208. *Eudoxus.*
                                                               207. *Aristoteles.*
                                                               197. *Meton.*
                                                               202. *Scoresby.*

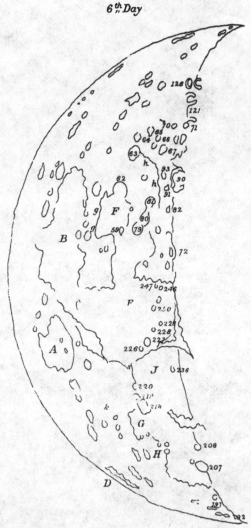

A   *Mare Crisium.*   F   *Mare Nectaris.*   g. *The Pyrenees.*
B   „   *Fœcunditatis.*   G   *Lacus Somniorum.*   h. *Altai Mts.*
D   „   *Humboldtianum.*   H   „   *Mortis.*   k. *Taurus Mts.*
E   „   *Tranquillitatis.*   J   *Mare Serenitatis.*   s. *Mt. Argæus.*

George Philip & Son, Ltd.

The London Geographical Institute

132. *Curtius.*

129. *Zach.*

128. *Lilius.*

124. *Licetus.*

123. *Clairaut.*

121. *Maurolycus.*

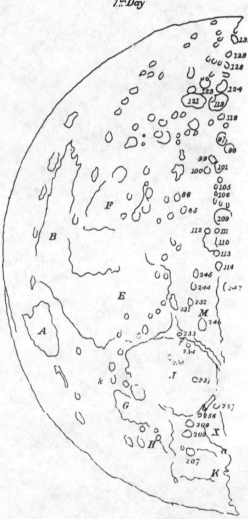

7.ᵗʰ Day

119. *Stöfler.*

118. *Fernelius.*

97. *Aliacensis.*

98. *Werner.*

99. *Apianus.*

100. *Playfair.*

101. *Blanchinus.*

105. *Donati.*

106. *Airy.*

86. *Almanon.*

85. *Abulfeda.*

109. *Albategnius.*

112. *Hind.*

111. *Halley.*

110. *Hipparchus.*

113. *Horrocks.*

114. *Rhaeticus.*

245. *Godin.*

244. *Agrippa.*

242. *Triesnecker.*

232. *Boscovich.*

231. *Julius-Cæsar.*

240. *Manilius.*

233. *Taquet.*

234. *Menelaus.*

236. *Bessel.*

237. *Linnee.*

257. *Theœtetus.*

256. *Calippus.*

209. *Alexander.*

208. *Eudoxus.*

207. *Aristoteles.*

A   *Mare Crisium.*

B   ,,   *Fœcunditatis.*

E   ,,   *Tranquillitatis.*

G   *Lacus Somniorum.*

F   *Mare Nectaris.*

K   ,,   *Frigoris.*

M   ,,   *Vaporum.*

J   ,,   *Serenitatis.*

H   *Lacus Mortis.*

X   *Palus Nebularum.*

M   *Mare Vaporum.*

b.   *The Caucasus.*

f.   *The Hæmus Mts.*

k.   *The Taurus Mts.*

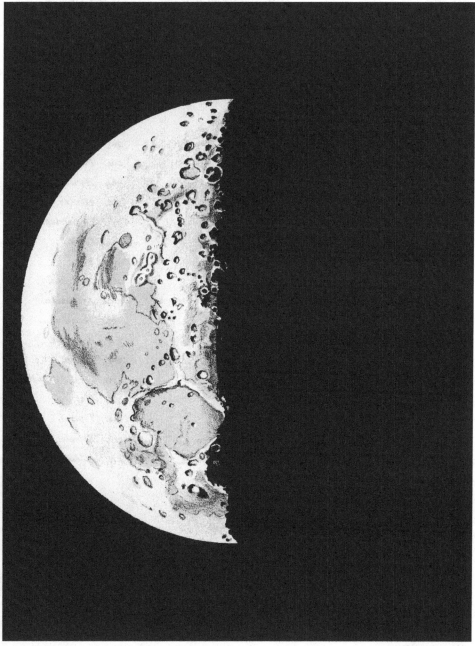

George Philip & Son. Ltd.

The London Geographical Institute

300. *Moretus.*

297. *Deluc.*

296. *Maginus.*

290. *Saussure.*

116. *Walter.*

285. *Ball.*

278. *Regiomontanus.*

277. *Purbach.*

274. *Thebit.*

266. *Arzachel.*

267. *Alpetragius.*

265. *Alphonsus.*

264. *Ptolemæus.*

263. *Herschel.*

261. *Mösting.*

373. *Sömmering.*

374. *Schröter.*

376. *Bode.*

260. *Autolycus.*

399. *Archimedes.*

257. *Theætetus.*

259. *Aristillus.*

258. *Cassini.*

211. *Great Alpine Valley.*

413. *Plato.*

414. *Timæus.*

416. *Epigenes.*

417. *Goldschmidt.*

8th Day

P   *Sinus Medii.*

M   ,,   *Æstuum.*

J   *Mare Serenitatis.*

L   ,,   *Imbrium.*

X   *Palus Nebularum.*

Z   ,,   *Putredinis.*

K   *Mare Frigoris.*

a.   *The Alps.*

b.   *The Caucasus.*

c.   *The Apennines.*

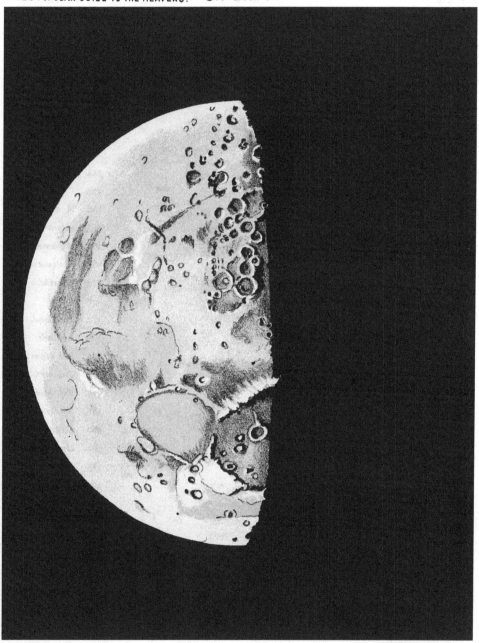

                                                          The London Geographical Institute

300. *Moretus.*   296. *Maginus.*   293. *Wilhelm*

299. *Cysatus.*   294. *Longomontanus.*   292. *Heinsius.*

298. *Clavius.*   291. *Tycho.*   282. *Gauricus.*

                                                          283. *Wurzelbauer.*

                                                          280. *Sasserides.*

                                                          281. *Hesiodus.*

                                                          275. *Straight Wall.*

                                                          344. *Nicollet.*

                                                          274. *Thebit.*

                                                          266. *Arzachel.*

                                                          267. *Alpetragius.*

                                                          265. *Alphonsus.*

                                                          264. *Ptolemæus.*

                                                          270. *Guerike.*

                                                          271. *Parry.*

                                                          272. *Bonpland.*

                                                          262. *Lalande.*

                                                          273. *Fra Mauro.*

                                                          372. *Gambart.*

                                                          377. *Reinhold.*

                                                          381. *Stadius.*

                                                          380. *Copernicus.*

                                                          382. *Eratosthenes.*

                                                          383. *Gay Lussac.*

                                                          403. *Pytheas.*

                                                          401. *Timocharis.*

                                                          402. *Lambert.*

                                                          399. *Archimedes.*

                                                          413. *Plato.*

                                                          419. *Fontenelle.*

                                                          417. *Goldschmidt.*

                                                          418. *Anaxagoras.*

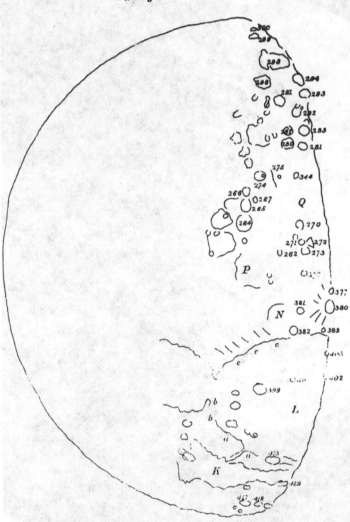

9ᵗʰ Day

K *Mare Frigoris.*   P *Sinus Medii.*   a. *The Alps.*

L   ,, *Imbrium.*   Q *Mare Nubium.*   b. *The Caucasus Alps.*

N *Sinus Æstuum.*   c. *The Apennines.*

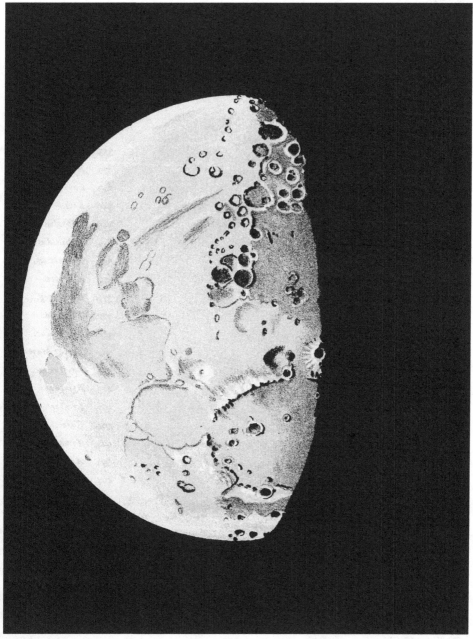

George Philip & Son, Ltd.

The London Geographical Institute

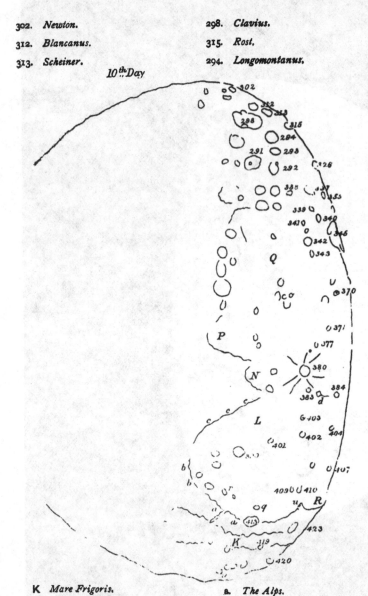

302. *Newton.*   298. *Clavius.*   291. *Tycho.*
312. *Blancanus.*   315. *Rost.*   293. *Wilhelm I.*
313. *Scheiner.*   294. *Longomontanus.*   292. *Heinsius.*

10th Day   326. *Hainzel.*
                                                 338. *Cichus.*
                                                 337. *Capuanus.*
                                                 355. *Ramsden.*
                                                 339. *Mercator.*
                                                 340. *Campanus.*
                                                 341. *Kies.*
                                                 345. *Hippalus.*
                                                 342. *Bullialdus.*
                                                 343. *Lubiniezky.*
                                                 370. *Euclides.*
                                                 371. *Landsberg.*
                                                 377. *Reinhold.*
                                                 380. *Copernicus.*
                                                 384. *Tobias Mayer.*
                                                 383. *Gay Lussac.*
                                                 403. *Pytheas.*
                                                 402. *Lambert.*
                                                 404. *Euler.*
                                                 401. *Timocharis.*
                                                 399. *Archimedes.*
                                                 407. *Caroline Herschel.*
                                                 409. *Leverrier.*
                                                 410. *Helicon.*
                                                 413. *Plato.*
                                                 423. *Condamine.*
                                                 419. *Fontenelle.*
                                                 420. *Philolaus.*

K   *Mare Frigoris.*   a. *The Alps.*
L   ,,   *Imbrium.*   b. *The Caucasus.*   r. *Piton.*
P   *Sinus Medii.*   c. *The Apennines.*   u. *Prom. Laplace.*
Q   *Mare Nubium.*   d. *The Carpathians.*
R   *Sinus Iridum.*   q. *Pico.*

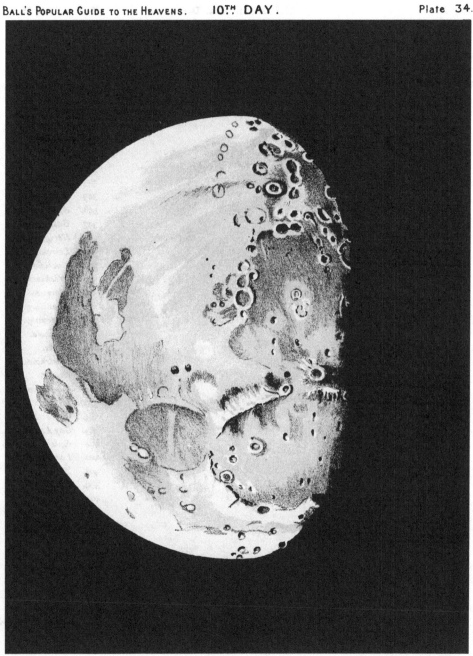

George Philip & Son, Ltd.   The London Geographical Institute

305. *Casatus.*
306. *Klaproth.*
307. *Wilson.*

II.th Day

309. *Bettinus.*
311. *Segner.*
313. *Scheiner.*

298. *Clavius.*
315. *Rost.*
317. *Schiller.*
318. *Bayer.*
326. *Hainzel.*
337. *Capuanus.*
355. *Ramsden.*
335. *Vitello.*
334. *Lee.*
333. *Doppelmayer.*
345. *Hippalus.*
346. *Agatharchides.*
347. *Gassendi.*
348. *Herigonius.*
349. *Letronne.*
370. *Euclides.*
369. *Wichmann.*
368. *Flamsteed.*
371. *Landsberg.*
385. *Kunowsky.*
377. *Reinhold.*
386. *Encke.*
378. *Hortensius.*
387. *Kepler.*
380. *Copernicus.*
379. *Milichius.*
384. *Tobias Mayer.*
383. *Gay Lussac.*
404. *Euler.*
402. *Lambert.*
401. *Timocharis.*
399. *Archimedes.*
405. *Diophantus.*
406. *Delisle.*
451. *Gruithuisen.*
427. *Mairan.*
426. *Sharp.*
425. *Bianchini.*
423. *Condamine.*
413. *Plato.*
429. *Harpalus.*
430. *J. F. W. Herschel.*
420. *Philolaus.*
421. *Anaximenes.*
i.   *Riphæan Mts.*
q.   *Pico.*
r.   *Piton Mountain.*
u.   *Prom, Laplace.*

L   *Mare Imbrium.*
K   ,, *Frigoris.*
Q   ,, *Nubium.*
R   *Sinus Iridum.*
S   *Oceanus Procellarum.*
T   *Mare Humorum.*

a.   *The Alps.*
b.   *The Caucasus.*
c.   *The Apennines.*
d.   *The Carpathians.*
e.   *The Sinus Iridum*
     Highlands.

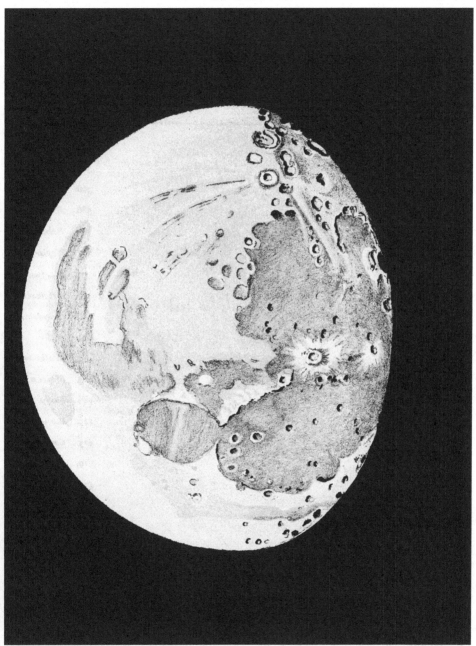

317. *Schiller.*
318. *Bayer.*
321. *Phocylides.*

12th Day

323. *Schichard.*
327. *Lehmann.*
332. *Vieta.*

351. *Cavendish.*
350. *Mersenius.*
347. *Gassendi.*
359. *Fontana.*
356. *Billy.*
357. *Hansteen.*
349. *Letronne.*
368. *Flamsteed.*
386. *Encke.*
389. *Reiner.*
387. *Kepler.*
390. *Marius.*
447. *Aristarchus.*
448. *Herodotus.*
404. *Euler.*
449. *Wollaston.*
427. *Mairan.*
426. *Sharp.*
425. *Bianchini.*
423. *Condamine.*
419. *Fontenelle.*
431. *Anaximander.*
420. *Philolaus.*
421. *Anaximenes.*

K   *Mare Frigoris.*
L   „   *Imbrium.*
Q   „   *Nubium.*

S   *Oceanus Procellarum.*
T   *Mare Humorum.*

e.   *The Sinus Iridum Highlands.*
u.   *Prom. Laplace.*

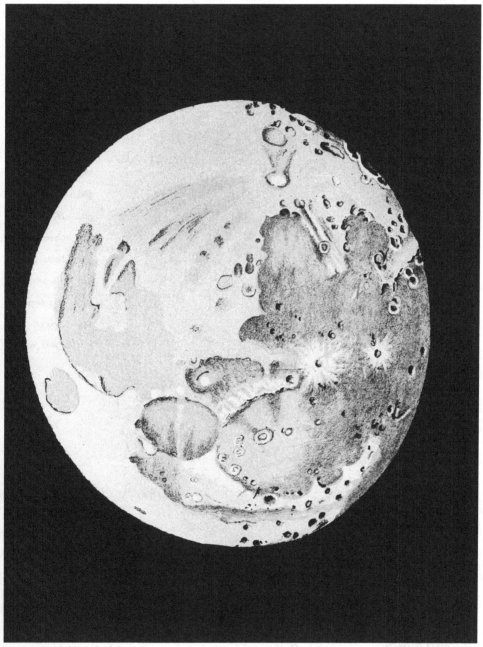

321. *Phocylides.*

322. *Wargentin.*

325. *Inghirami.*

323 *Schickard.*

329. *Piazzi.*

330. *Lagrange.*

332. *Vieta.*

351. *Cavendish.*

350. *Mersenius.*

359. *Fontana.*

361. *Crüger.*

362. *Rocca.*

363. *Grimaldi.*

366. *Lohrmann.*

391. *Hevel.*

392. *Cavalerius.*

397. *Seleucus.*

448. *Herodotus.*

447. *Aristarchus.*

445. *Briggs.*

432. *Pythagoras.*

13ᵗʰ Day.

L   *Mare Imbrium.*

Q   ,,   *Nubium.*

R   *Sinus Iridum.*

S   *Oceanus Procellarum.*

T   *Mare Humorum.*

W   *Sinus Roris.*

e.   *The Sinus Iridum Highlands.*

u.   *Prom. Laplace.*

316. *Bailly.*

321. *Phocylides.*

322. *Wargentin.*

14ᵗʰ Day

325. *Inghirami.*

323. *Schickard.*

329. *Piazzi.*

330. *Lagrange.*

353. *Eichstädt.*

362. *Rocca.*

363. *Grimaldi.*

365. *Riccioli.*

366. *Lohrmann.*

391. *Hevel.*

392. *Cavalerius.*

393. *Olbers.*

394. *Cardanus.*

395. *Kraft.*

448. *Herodotus.*

447. *Aristarchus.*

397. *Seleucus.*

445. *Briggs.*

446. *Otto Struve.*

439. *Repsold.*

438. *Xenophanes.*

437. *Cleostratus.*

432. *Pythagoras.*

K   *Mare Frigoris.*

L   *Mare Imbrium.*

R   *Sinus Iridum.*

S   *Oceanus Procellarum.*

T   *Mare Humorum.*

W   *Sinus Roris.*

CHAPTER VI.—THE SKY MONTH BY MONTH, AND THE INDEX TO THE PLANETS.

PLATES 39—50.

THE MONTHLY MAPS.

The diurnal rotation of the earth gives rise to an apparent revolution of the celestial sphere in a period of one sidereal day. In consequence of this movement the appearance of the sky is continually changing, so that to the beginner it is often a matter of considerable difficulty to know where to look for any particular star or constellation.

As the sidereal day is about 4 minutes shorter than the ordinary mean solar day, the effect produced is a gradual shifting of the stars from east to west,—a star which occupies a certain position one night, reaches the same position 4 minutes earlier the next night, so that at the end of a month this position is attained 2 hours earlier than at the beginning.

In order to render these changes easier to follow, and to enable the student to identify the principal constellations without difficulty, and to know where any particular star or group of stars is to be found at any time, Plates 39—50 are used. They represent the positions of the principal stars down to the 4th magnitude at intervals of 2 sidereal hours. The first shows the aspect of the heavens at midnight on January 15th, the sidereal time then being 7h. 37m. This map will also represent the appearance of the visible hemisphere at the times shown in the corners at the top. Thus we find that the first of the monthly maps may be used in February at 10 P.M., in March at 8 P.M. From April to September inclusive, the stars will occupy the positions here indicated during the daylight hours, when they will be invisible ; but in October this aspect of the sky may again be seen at 6 A.M., in November at 4 A.M., and in December at 2 A.M. To find the right map for any month and hour we can make use of the following Table.

TABLE TO FIND THE ASPECT OF THE HEAVENS AT ANY GIVEN MONTH AND HOUR OF NIGHT.

	P.M. 4h.	P.M. 6h.	P.M. 8h.	P.M. 10h.	Mid-night. 12h.	A.M. 2h.	A.M. 4h.	A.M. 6h.	A.M. 8h.
January....	47	48	49	50	39	40	41	42	43
February ..		49	50	39	40	41	42	43	
March		50	39	40	41	42	43	44	
April			40	41	42	43	44		
May			41	42	43	44	45		
June			42	43	44	45	46		
July.........			43	44	45	46	47		
August			44	45	46	47	48		
September.		44	45	46	47	48	49	50	
October.....		45	46	47	48	49	50	39	
November..	45	46	47	48	49	50	39	40	41
December..	46	47	48	49	50	39	40	41	42

EXAMPLES OF THE USE OF THIS TABLE :

I. To find the map suitable for 10 P.M. in March. Take the third row, and under 10 P.M is found 40. This means Plate 40.

II. What map should be used at 7-30 P.M. in November ? On the eleventh row we find 47 under 8 P.M., and as 7-30 is nearer to 8 than to 6, we accordingly choose Plate 47.

It will of course be understood that the maps have been designed to represent the appearance of the sky on the 15th of each month, at the hours mentioned. The changes are, however, so slow, that for most purposes they will be found sufficiently applicable to the whole month. If, however, greater precision is desired, it can be obtained by *subtracting* half-an-hour from the time given on the map for each week *after*, or *adding* half-an-hour for each week *before* the middle of the month. Plate 39 is thus quite accurate at 11 P.M. on the 30th of January, or at 8-30 P.M. on March 8th ; similarly Plate 41 is correct at 5 A.M. on the 30th December, or at 11 P.M. on the 1st April.

These maps have been constructed for the latitude 53° 23′ N. They will thus be suitable for all parts of the British Isles. The bounding circle represents the horizon, and the small cross at the centre marks the position of the zenith. The projection used is such that the distance of each star from the centre of the map is proportional to the star's zenith distance. The angle which the line joining a star to the zenith makes with the central meridian is the azimuth of the star.

As the celestial sphere is viewed from the inside, the cardinal points are not disposed on these maps as in a terrestrial atlas. In the present case, when the north is at the top, the west will be on the right, and the east on the left.

If we wish to compare any region of the sky with the map, we suppose a radius drawn through the middle of this region, and the point where it cuts the circumference of the map gives us the azimuth. Turning our face towards this point of the compass, we hold the map so that the corresponding point of the circumference is lowest, and, remembering that the centre of the circle represents the zenith, we have on the map a picture of the corresponding position of the sky. For instance, if we wish to find the constellation Leo at midnight, in the middle of March, we find from Map 41 that the radius drawn through the middle of it cuts the circumference at about ⅔ths of the way from south towards west. We accordingly turn to that point of the horizon, and we can readily find this constellation. The brightest star, Regulus, will be then almost exactly half-way between the zenith and horizon, or at an altitude of 45°, while the "Sickle," which forms the fore-part of this constellation, will be found tilted over towards the west. If we turn a little further towards the west, we shall find the two bright stars, Castor and Pollux, a little lower in the sky, the line joining them being nearly horizontal. Again, if in October, at 10 P.M., we wish to examine "The Plough," as the group formed by the principal stars in Ursa Major are called, we go to Plate 47 and find this group almost due north. We have accordingly to turn the map upside down, so as to bring the north point lowest, and we then see this figure stretching across the sky in a horizontal direction, at an altitude of about 20°. If we turn to the north-east, we again find Castor and Pollux just above the horizon, but this time the line joining them is very nearly vertical.

The names of the constellations have been printed on the maps, so that when the maps are held in the proper position for any constellation its name may be erect.

As the student becomes more familiar with the stars, he will probably wish to identify many fainter groups that do not appear on these plates. In order to enable this to be done with facility, the faint dotted lines have been inserted which mark the boundaries of the regions which each of the plates, 51—70, of the general Atlas, cover on the sky. These lines appear everywhere in pairs, the spaces between the pairs being the areas by which the maps overlap each other. The numbers within the regions thus marked out are those of the corresponding plates in this volume, where more detailed maps of the same part of the sky will be found. Thus in Plate 40 we find the constellation Leo almost wholly contained in the space corresponding to Plate 60, and some of its principal stars in the space common to Plates 54, 55, and 60. If we turn to Plate 60 we shall find the whole group on a much larger scale, while 54 and 55 show the more northern parts of this constellation.

THE INDEX TO THE PLANETS.

It is a special object of this work to indicate from month to month and from year to year the positions in the sky of the principal planets. Let it be once for all understood that those who want *exact* positions must seek for them elsewhere—in the *Nautical Almanac* for example. What is here given is an index to the planets sufficient for the following purposes :

(1) To find the place in the heavens which each principal planet occupies in each month of the years 1901—1950.
(2) To find when any principal planet rises, or souths, or sets.
(3) To determine thence the best season of any year for the observation of the planet.
(4) To find the name of a planet when the time and place of its appearance are known.

The following table gives the dates of the Planetary Phenomena described at the head of each column.

PLANETARY PHENOMENA.

A.D.	Mercury.		Venus. Greatest Elongation.		Mars in Opposition.	Jupiter in Opposition.	Saturn in Opposition.	Phase of Saturn Plate 8
	Evening Star.	Morning Star.	Evening.	Morning.				
1901	Feb. 16—23	July 29—Aug. 5	December	—	February	June	July	7
1902	Jan. 30—Feb. 6	Oct. 31—Nov. 7	—	April	—	August	July	7
1903	May 5—12	Oct. 15—22	July	November	March	September	July	8
1904	Apr. 18—25	Sept. 28—Oct. 5	—	—	—	October	August	8
1905	Mar. 30—Apr. 6	Sept. 11—18	February	July	May	November	August	8
1906	Mar. 14—21	Aug. 26—Sept. 2	September	—	—	December	September	9
1907	Feb. 26—Mar. 5	Aug. 9—16	—	February	June	—	September	9
1908	Feb. 9—16	Nov. 10—17	April	September	—	January	September	9
1909	Jan. 22—29	Oct. 25—Nov. 1	December	—	September	February	October	10
1910	Apr. 28—May 5	Oct. 8—15	—	April	—	March	October	10
1911	Apr. 11—18	Sept. 21—28	July	November	November	May	November	11
1912	Mar. 25—Apr. 1	Sept. 5—12	—	—	—	June	November	11
1913	Mar. 8—15	Aug. 19—26	February	July	—	July	December	12
1914	Feb. 19—26	Aug. 1—8	September	—	January	August	December	12
1915	Feb. 2—9	Nov. 3—10	—	February	—	September	December	12
1916	May 9—16	Oct. 18—25	April	September	February	October	—	1
1917	Apr. 21—28	Oct. 1—8	November	—	—	November	January	1
1918	Apr. 3—10	Sept. 14—21	—	April	March	—	January	1
1919	Mar. 17—24	Aug. 29—Sept. 5	July	November	—	January	February	2
1920	Mar. 1—8	Aug. 12—19	—	—	April	February	February	2
1921	Feb. 12—19	Nov. 13—20	February	July	—	March	March	3
1922	Jan. 25—Feb. 1	Oct. 28—Nov. 4	September	—	June	April	March	3
1923	May 1—8	Oct. 11—18	—	February	—	May	April	4
1924	Apr. 14—21	Sept. 24—Oct. 1	April	September	August	June	April	4
1925	Mar. 28—Apr. 4	Sept. 8—15	November	—	—	July	April	4
1926	Mar. 11—18	Aug. 22—29	—	April	October	August	May	5
1927	Feb. 22—Mar. 1	Aug. 4—11	July	November	—	September	May	5
1928	Feb. 5—12	Nov. 6—13.	—	—	December	October	June	6
1929	May 12—19	Oct. 21—28	February	June	—	December	June	6
1930	Apr. 24—May 1	Oct. 4—11	September	—	—	—	June	6
1931	Apr. 6—13	Sept. 17—24	—	February	January	January	July	7
1932	Mar. 20—27	Sept. 1—8	April	September	—	February	July	7
1933	Mar. 4—11	Aug. 15—22	November	—	March	March	August	8
1934	Feb. 15—22	July 29—Aug. 5	—	April	—	April	August	8
1935	Jan. 28—Feb. 4	Oct. 31—Nov. 7	July	November	April	May	August	8
1936	May 4—11	Oct. 14—21	—	—	—	June	September	9
1937	Apr. 17—24	Sept. 27—Oct. 4	February	June	May	July	September	9
1938	Mar. 31—Apr. 7	Sept. 11—18	September	—	—	August	October	10
1939	Mar. 14—21	Aug. 25—Sept. 1	—	February	July	September	October	10
1940	Feb. 25—Mar. 3	Aug. 7—14	April	September	—	November	November	11
1941	Feb. 8—15	Nov. 9—16	November	—	September	December	November	11
1942	Jan. 22—29	Oct. 24—31	—	April	—	—	November	11
1943	Apr. 27—May 4	Oct. 7—14	June	November	December	January	December	12
1944	Apr. 9—16	Sept. 20—27	—	—	—	February	December	12
1945	Mar. 23—30	Sept. 4—11	February	June	—	March	—	1
1946	Mar. 7—14	Aug. 18—25	September	—	January	April	January	1
1947	Feb. 18—25	Aug. 1—8	—	January	—	May	January	1
1948	Feb. 1—8	Nov. 3—10	April	September	February	June	February	2
1949	May 7—14	Oct. 17—24	November	—	—	July	February	2
1950	Apr. 20—27	Sept. 30—Oct. 7	—	April	March	August	March	3

Since Mercury and Venus move in orbits lying between the Earth and the Sun, they appear from the Earth to be always comparatively near the Sun, and are visible consequently only in the hours succeeding sunset or preceding sunrise. On the average they are most favourably seen at the times when their angular distance from the Sun is greatest—the times of greatest elongation. But the circumstances are greatly modified by the season of the year at which these times fall. The planets move nearly in the Ecliptic. In the spring, even in latitudes as far north as England, at sunset the Ecliptic rises steeply above the western horizon, and the planets are correspondingly high for a given elongation from the Sun. In autumn, at sunset, the Ecliptic lies low along the south-western horizon, and the planets are correspondingly low.

MERCURY.

It follows from what has just been said that Mercury is best seen after sunset, when it comes to greatest elongation east of the Sun in the spring, and best before sunrise when it comes to greatest elongation west in the autumn. We have given, therefore, in the table of Planetary Phenomena for Mercury, the week about the dates of greatest elongations east and west, which come respectively in spring and autumn ; and these are in general the most favourable times of year for finding the planet. Owing, however, to the considerable inclination, 7°, of the planet's orbit, some spring and autumn elongations are much more favourable than others, and exact details for each year must be obtained from the *Nautical Almanac.*

Mercury moves so quickly among the stars, completing a circuit from conjunction to conjunction in 116 days, that it is not possible within the limits of this work to give an index to Mercury, as is done for the other principal planets.

VENUS.

The fourth and fifth columns of the Table of Planetary Phenomena give the month in which Venus attains her greatest elongation east and west of the Sun. At these times she is at her greatest angular distance from the Sun, and there is then on the average the greatest chance of seeing her well. But the advantages in this respect of different elongations are profoundly modified by the time of year at which they occur. For some months before the time of greatest eastern elongation Venus is well to the east of the Sun, and if it happens that the time of year is spring when the Ecliptic rises steeply above the western horizon, she will be a long way north as well as east of the Sun, will set all the later in consequence, and will be seen under more favourable conditions, though not so bright as when she reaches greatest elongation, a month before she is most brilliant. Taking everything into account, in northern latitudes Venus is at her best as an evening star when she comes to greatest elongation east in June or July. She is then conspicuous after sunset all through the spring months, but is low and not so well seen by the time she reaches greatest brightness. If, on the other hand, she comes to greatest elongation east in February, and to greatest brightness in March, she makes a less prolonged but more splendid appearance. Bearing in mind that the factors to be taken into account are (1) elongation, (2) distance north of the Sun, (3) brightness (greatest a month *before* western elongation), it will not be difficult to frame similar rules for her appearances as a morning star.

The Index to Venus, by which title we have designated the following table, enables the approximate position of the planet to be readily ascertained for any month up to the end of 1950. At the top of the index the names of the months are given in a horizontal row. The first column gives the year. Corresponding to each month of each year the index shows a number, which is the number of one of the monthly maps, 39—50. The planet will be found in the region of the sky where the central meridian of the map cuts the "Track of the Planets."

It must be remembered that, as the unit of time adopted in this index, as well as in those of the other planets, is a month, and as the locality can only be indicated by dividing the "Track of the Planets," around the heavens into twelve portions, no close precision need be looked for. No doubt in the great majority of cases the map named in the index will be that where the central meridian lies nearest the planet. In other words, the place of the planet is given to within an hour. It may, however, happen in extreme conditions that the map indicated is not the best one, but in such cases the right map always lies next to that to which the index refers. Even when this happens, the purposes of the index are not frustrated, for the planet will lie so near to the position in question, that its identification will be unmistakable, unless on the rare occasions when two planets happen to lie close together.

EXAMPLES TO ILLUSTRATE THE USE OF THE INDEX TO VENUS.

Example 1.—What will be the aspect of Venus in March, 1908 ?

Solution.—The index given for March, 1908, the number 49. On this Plate the
central meridian cuts the "Track of the Planets" in Taurus. This is the region
of the sky required. The next question is, when is this region above the horizon
in England ? The Table at the top of Plate 49 shows that it is on the meridian
about 4 p.m. in March. It will be well above the western horizon, therefore, at
sunset, and Venus will be splendidly conspicuous.

Example 2.—Where will Venus be when brightest as an evening star in 1943 ?

Solution.—The Table of Planetary Phenomena shows that in 1943 Venus is at
greatest elongation east in June. She will, therefore, be brightest in July. The
index gives for 1943, July, the number 40, and the corresponding chart shows
that Venus is in Leo. Turning forward to plates 43 and 44 we see that at
about 10 p.m. in July, Venus will be setting in the N.W., and as the twilight
will be strong the planet will not appear to great advantage.

INDEX TO VENUS.

A.D.	Jan.	Feb.	March	April	May	June	July	August	Sept.	Oct.	Nov.	Dec.
1901	44	46	47	48	49	39	40	41	42	43	45	46
1902	46	46	46	47	48	49	50	39	40	41	43	44
1903	45	47	48	49	50	40	41	41	41	41	41	42
1904	44	45	46	47	49	50	39	40	42	43	44	45
1905	47	48	48	49	48	49	49	39	40	41	42	43
1906	45	46	47	49	50	39	40	42	43	43	43	43
1907	44	45	46	47	48	49	50	40	41	42	44	45
1908	46	47	49	50	50	50	50	39	40	41	42	43
1909	44	46	47	48	49	39	40	41	42	43	45	46
1910	46	46	46	47	48	49	50	39	40	41	43	44
1911	45	47	48	49	50	40	41	41	41	41	41	42
1912	44	45	46	47	49	50	39	40	42	43	44	45
1913	47	48	48	49	48	49	49	39	40	41	42	43
1914	45	46	47	49	50	39	40	42	43	43	43	43
1915	44	45	46	47	48	49	50	40	41	42	44	45
1916	46	47	49	50	50	50	50	39	40	41	42	43
1917	44	46	47	48	49	39	40	41	42	44	45	45
1918	45	45	46	47	48	49	50	39	40	42	43	44
1919	46	47	48	49	50	40	41	41	41	41	41	42
1920	44	45	46	47	49	50	39	40	42	43	44	45
1921	47	48	48	49	48	49	49	39	40	41	42	43
1922	45	46	47	49	50	39	40	42	43	43	43	43
1923	44	45	46	47	48	49	50	40	41	42	44	45
1924	46	47	49	50	50	50	50	39	40	41	42	43
1925	44	46	47	48	49	39	40	41	42	44	45	45
1926	45	45	46	47	48	49	50	39	40	42	43	44
1927	46	47	48	49	50	40	41	41	41	41	41	42
1928	44	45	46	47	49	50	39	40	42	43	44	45
1929	47	48	48	49	48	49	49	39	40	41	42	43
1930	45	46	47	49	50	39	40	42	43	43	43	43
1931	44	45	46	47	48	49	50	40	41	42	44	45
1932	46	47	49	50	50	50	50	39	40	41	42	43
1933	44	46	47	48	49	39	40	41	42	44	45	45
1934	45	45	46	47	48	49	50	39	40	42	43	44
1935	46	47	48	49	50	40	41	41	41	41	41	42
1936	44	45	46	47	49	50	39	40	42	43	44	45
1937	47	48	48	49	48	49	49	39	40	41	42	43
1938	45	46	47	49	50	39	40	42	43	43	43	43
1939	44	45	46	47	48	49	50	40	41	42	44	45
1940	46	47	49	50	50	50	50	39	40	41	42	43
1941	44	46	47	48	49	39	40	41	42	44	45	45
1942	45	45	46	47	48	49	50	39	40	42	43	44
1943	46	47	48	49	39	40	40	40	40	40	41	43
1944	44	45	46	48	49	50	39	41	42	43	44	45
1945	47	48	48	49	48	49	49	39	40	41	42	43
1946	45	46	47	49	50	39	40	42	43	43	43	43
1947	44	45	46	47	48	49	50	40	41	42	44	45
1948	46	47	49	50	50	50	50	39	40	41	42	43
1949	44	46	47	48	49	39	40	41	42	44	45	45
1950	45	45	46	47	48	49	50	39	40	42	43	44

MARS.

We have already explained, in connection with Plate 6, the phases of Mars and the effect of the eccentricity of his orbit, whereby the distance of the planet, and its consequent apparent diameter, is dependent upon the time of year at which opposition occurs—the most favourable date being the end of August, and the least favourable the end of February. It should not be forgotten, however, that in countries as far north as England, Mars at an August opposition is very low down in the south, and unfavourably placed for telescopic scrutiny. It may well happen that at an opposition occuring later in the year, say in October, the planetary detail is better seen, because of the greater elevation at which the planet crosses the meridian, although the planet is farther away and apparently smaller.

The conditions for a favourable view of Mars and other exterior planets are much simpler than those for Venus and Mercury. The planet is best seen when it is in opposition to the Sun ; it is then on the meridian at midnight, and is up for practically the whole of the night.

The table of planetary phenomena gives the dates of the opposition of Mars, and we can judge from them at once how favourable the opposition will be. The Index to Mars enables us to find in what region of the sky the planet will be at any time.

EXAMPLES TO ILLUSTRATE THE USE OF THE INDEX TO MARS.

Example 1.—Find the place of Mars in August, 1907.

Solution.—The Index gives the reference 45, and the corresponding chart shows that the planet lies between Capricornus and Sagittarius. It is on the meridian about 10 p.m., but is considerably south. A reference to the Table of Planetary Phenomena shows that Opposition occurred in June, which is not a very favourable case, and, since the planet is also far south, it will not be very favourably placed for observation in England in 1907, August.

Example 2.—When is Mars on the meridian in April, 1938 ?

Solution.—The index gives 48, and the corresponding map shows that Mars is in Aries. Reference to the table in the top corner shows that Aries is on the meridian at noon in April, during daylight ; the planet will, therefore, not be visible.

INDEX TO MARS.

A.D.	Jan.	Feb.	March	April	May	June	July	August	Sept.	Oct.	Nov.	Dec.
1901	41	40	40	40	40	41	41	42	43	43	44	45
1902	46	46	47	48	49	49	50	39	40	40	41	41
1903	42	42	41	41	41	41	42	42	43	44	44	45
1904	46	47	47	48	49	50	50	39	40	40	41	42
1905	42	42	43	43	43	43	43	43	44	44	45	46
1906	47	47	48	49	50	50	39	40	40	41	42	42
1907	43	44	44	45	45	44	45	45	45	46	46	47
1908	48	48	49	50	50	39	40	40	41	42	42	43
1909	44	45	45	46	47	47	47	47	47	47	47	47
1910	48	48	49	50	50	39	40	40	41	42	43	43
1911	44	44	45	46	47	47	48	49	49	49	49	49
1912	49	49	50	50	39	40	40	41	41	42	43	43
1913	44	45	46	47	47	48	49	49	50	39	39	39
1914	39	39	39	39	40	40	41	42	42	43	44	44
1915	45	46	47	48	48	49	50	39	39	40	40	41
1916	41	40	40	40	40	41	41	42	43	43	44	45
1917	46	46	47	48	49	49	50	39	40	40	41	41
1918	42	42	41	41	41	41	42	42	43	44	44	45
1919	46	47	47	48	49	50	50	39	40	40	41	42
1920	42	42	43	42	42	42	42	43	43	44	45	46
1921	47	47	48	49	49	50	39	40	40	41	41	42
1922	43	43	44	44	44	44	44	44	44	45	46	46
1923	47	48	48	49	50	39	39	40	41	41	42	42
1924	43	44	45	45	46	46	46	46	46	46	46	47
1925	47	48	49	49	50	39	39	40	41	41	42	43
1926	43	44	45	46	46	47	48	48	48	48	48	48
1927	48	49	49	50	39	39	40	41	41	42	43	43
1928	44	45	45	46	47	48	48	49	50	50	50	50
1929	50	50	50	39	39	40	41	41	42	42	43	44
1930	45	46	46	47	48	49	49	50	39	39	40	40
1931	39	39	39	39	40	40	41	42	42	43	44	44
1932	45	46	47	48	48	49	50	39	39	40	40	41
1933	41	41	41	41	41	41	42	42	43	44	44	45
1934	46	47	47	48	49	50	50	39	40	40	41	42
1935	42	42	43	42	42	42	42	43	43	44	45	46
1936	47	47	48	49	49	50	39	40	40	41	41	42
1937	43	43	44	44	43	43	43	43	44	44	45	46
1938	47	47	48	49	50	50	39	40	40	41	42	42
1939	43	44	44	45	45	45	45	45	45	46	46	46
1940	47	48	48	49	50	50	39	40	40	41	42	42
1941	43	44	45	45	46	47	47	47	47	47	47	47
1942	48	48	49	50	50	39	40	40	41	42	42	43
1943	44	44	45	46	47	47	48	49	49	50	50	50
1944	50	50	50	39	39	40	41	41	42	42	43	44
1945	45	46	46	47	48	49	49	50	39	39	40	40
1946	39	39	39	39	40	40	41	42	42	43	44	44
1947	45	46	47	48	48	49	50	39	39	40	40	41
1948	41	40	40	40	40	41	41	42	43	43	44	45
1949	46	46	47	48	49	49	50	39	40	40	41	41
1950	42	42	41	41	41	41	42	42	43	44	44	45

JUPITER.

Since the eccentricity of Jupiter's orbit is not large, and it lies far outside the orbit of the Earth, successive oppositions of the planet, every thirteen months, would be almost equally favourable, were it not for the fact that at a summer opposition the planet is always low in England, whereas at a winter opposition it is high. The dates of opposition from 1901—1950 are found in the Table of Planetary Phenomena. The position of the planet in the sky at any time can be found as usual from the following index.

EXAMPLES TO ILLUSTRATE THE USE OF THE INDEX TO JUPITER.

Example I.—In what part of the heavens will Jupiter be in January, 1910?

Solution.—The index to Jupiter gives for this month the number 41, and the corresponding Plate shows that the planet is between Leo and Virgo, and consequently close to the equator. The Table at the top shows that this region is on the meridian about 4 a.m. in January. The planet will therefore rise about 10 p.m., and will be well visible after midnight.

Example 2.—In July, 1919, a bright planet is seen in the west at sunset. Is it Jupiter?

Solution.—The index to Jupiter gives for July, 1919, the reference to chart 10, which shows that Jupiter is a little to the west of the Sickle in Leo. Turning on to Plates 43 and 44, we see that the Sickle is low down at 8 p.m., and is on the horizon at 10 p.m. Jupiter will therefore be scarcely visible at Sunset, and the planet in question is probably not Jupiter. Reference to the Index to Venus shows that the planet higher up in the west at that time is Venus.

INDEX TO JUPITER.

A.D.	Jan.	Feb.	March	April	May	June	July	August	Sept.	Oct.	Nov.	Dec.
1901	44	44	44	44	44	44	44	44	44	44	44	45
1902	45	45	45	45	45	45	45	46	46	46	46	46
1903	46	47	47	47	47	47	47	47	47	47	47	47
1904	47	47	48	48	48	48	48	48	48	48	48	48
1905	48	48	48	49	49	49	49	49	49	49	49	49
1906	49	49	49	49	50	50	50	50	50	50	50	50
1907	50	50	50	50	50	39	39	39	39	39	39	39
1908	39	39	39	39	39	39	40	40	40	40	40	40
1909	40	40	40	40	40	40	40	41	41	41	41	41
1910	41	41	41	41	41	41	41	41	42	42	42	42
1911	42	42	42	42	43	43	43	43	43	43	44	44
1912	44	44	44	44	44	44	44	44	44	44	44	45
1913	45	45	45	45	45	45	45	45	45	45	45	45
1914	46	46	46	46	46	46	46	46	46	46	46	46
1915	46	47	47	47	47	47	47	47	47	47	47	47
1916	47	47	48	48	48	48	48	48	48	48	48	48
1917	48	48	48	49	49	49	49	49	49	49	49	49
1918	49	49	49	49	50	50	50	50	50	50	50	50
1919	39	39	39	39	39	39	40	40	40	40	40	40
1920	40	40	40	40	40	40	40	41	41	41	41	41
1921	41	41	41	41	41	41	41	41	42	42	42	42
1922	42	42	42	42	42	42	42	42	42	43	43	43
1923	43	43	43	43	43	43	43	43	43	43	44	44
1924	44	44	44	44	44	44	44	44	44	44	44	45
1925	45	45	45	45	45	45	45	45	45	45	45	45
1926	46	46	46	46	46	46	46	46	46	46	46	46
1927	46	47	47	47	47	47	47	47	47	47	47	47
1928	47	47	48	48	48	48	48	48	48	48	48	48
1929	48	48	48	49	49	49	49	49	49	49	49	50
1930	50	50	50	50	50	39	39	39	39	39	39	39
1931	39	39	39	39	39	39	40	40	40	40	40	40
1932	40	40	40	40	40	40	40	41	41	41	41	41
1933	41	41	41	41	41	41	41	41	42	42	42	42
1934	42	42	42	42	42	42	42	42	42	43	43	43
1935	43	43	43	43	43	43	43	43	43	43	44	44
1936	44	44	44	44	44	44	44	44	44	44	44	45
1937	45	45	45	45	45	45	45	45	45	45	45	46
1938	46	46	46	46	46	46	46	46	46	46	46	46
1939	46	47	47	47	47	47	47	47	47	47	47	47
1940	47	47	48	48	48	48	48	48	48	48	49	49
1941	49	49	49	49	50	50	50	50	50	50	50	50
1942	50	50	50	50	50	39	39	39	39	39	39	39
1943	39	39	39	39	39	39	40	40	40	40	40	40
1944	40	40	40	40	40	40	40	41	41	41	41	41
1945	41	41	41	41	41	41	41	41	42	42	42	42
1946	42	42	42	42	42	42	42	42	42	43	43	43
1947	43	43	43	43	43	43	43	43	43	43	44	44
1948	44	44	44	44	44	44	44	44	44	44	44	45
1949	45	45	45	45	45	45	45	45	45	45	45	45
1950	46	46	46	46	46	46	46	46	46	46	46	46

SATURN.

What we have said of Jupiter applies equally well to Saturn. An opposition is favourable or unfavourable according as the planet is in the northern or southern part of the ecliptic. But there is the further consideration, that in the telescope a great deal, and to the naked eye something, depends upon whether the rings are seen well open or closed up. When the rings are fully open the planet is nearly a magnitude brighter than when they are edgewise. For the appearance of the rings at any time reference may be made from the last column of the Table of Planetary Phenomena to Plate 6.

EXAMPLES TO ILLUSTRATE THE USE OF THE INDEX TO SATURN.

Example 1.—When will Saturn be found near the Pleiades ?

Solution.—The Pleiades are close to the central meridian of Chart 49.   The index gives the reference number, 49, for Saturn from May, 1911 to May, 1913. During these two years the planet will pass from west to east across the section of the "track of the planets" central on Chart 49, and will be close to the Pleiades in the summer of 1912.   The last column of the Table of Planetary Phenomena gives the number 11, and reference to Plate 6 shows that the rings are nearly at their widest open.   A repetition of the circumstance occurs in 1941 and 1942.

Example 2.—When will Saturn be found near the meridian at 10 p.m. in February ?

Solution.—Reference to the monthly maps shows that the region of the sky which comes to the meridian at 10 p.m. in February is on the meridian at midnight in January, and the Index number is 39.   We therefore look for this number in the Index to Saturn in the column for February, and find the years 1917, 1918, 1946, 1947.   On these years the planet will be on the meridian about 10 p.m. in February.

INDEX TO SATURN.

A.D.	Jan.	Feb.	March	April	May	June	July	August	Sept.	Oct.	Nov.	Dec.
1901	45	45	45	45	45	45	45	45	45	45	45	45
1902	45	45	45	45	45	45	45	45	45	45	45	45
1903	45	45	45	45	45	45	45	45	45	45	45	45
1904	45	46	46	46	46	46	46	46	46	46	46	46
1905	46	46	46	46	46	46	46	46	46	46	46	46
1906	46	46	47	47	47	47	47	47	47	47	47	47
1907	47	47	47	47	47	47	47	47	47	47	47	47
1908	47	47	47	47	47	47	47	47	47	47	47	47
1909	47	47	47	48	48	48	48	48	48	48	48	48
1910	48	48	48	48	48	48	48	48	48	48	48	48
1911	48	48	48	48	49	49	49	49	49	49	49	49
1912	49	49	49	49	49	49	49	49	49	49	49	49
1913	49	49	49	49	49	50	50	50	50	50	50	50
1914	50	50	50	50	50	50	50	50	50	50	50	50
1915	50	50	50	50	50	50	50	50	50	50	50	50
1916	50	50	50	50	50	50	39	39	39	39	39	39
1917	39	39	39	39	39	39	39	39	39	39	39	39
1918	39	39	39	39	39	39	39	40	40	40	40	40
1919	40	40	40	40	40	40	40	40	40	40	40	40
1920	40	40	40	40	40	40	40	40	41	41	41	41
1921	41	41	41	41	41	41	41	41	41	41	41	41
1922	41	41	41	41	41	41	41	41	41	42	42	42
1923	42	42	42	42	42	42	42	42	42	42	42	42
1924	42	42	42	42	42	42	42	42	42	42	42	42
1925	42	42	42	42	42	42	42	42	42	42	43	43
1926	43	43	43	43	43	43	43	43	43	43	43	43
1927	43	43	43	43	43	43	43	43	43	43	43	44
1928	44	44	44	44	44	44	44	44	44	44	44	44
1929	44	44	44	44	44	44	44	44	44	44	44	44
1930	44	44	44	44	44	44	44	44	44	44	44	44
1931	45	45	45	45	45	45	45	45	45	45	45	45
1932	45	45	45	45	45	45	45	45	45	45	45	45
1933	45	46	46	46	46	46	46	46	46	46	46	46
1934	46	46	46	46	46	46	46	46	46	46	46	46
1935	46	46	46	46	46	46	46	46	46	46	46	46
1936	46	46	47	47	47	47	47	47	47	47	47	47
1937	47	47	47	47	47	47	47	47	47	47	47	47
1938	47	47	47	48	48	48	48	48	48	48	48	48
1939	48	48	48	48	48	48	48	48	48	48	48	48
1940	48	48	48	48	49	49	49	49	49	49	49	49
1941	49	49	49	49	49	49	49	49	49	49	49	49
1942	49	49	49	49	49	49	49	49	49	49	49	49
1943	49	49	49	49	49	50	50	50	50	50	50	50
1944	50	50	50	50	50	50	50	50	50	50	50	50
1945	50	50	50	50	50	50	39	39	39	39	39	39
1946	39	39	39	39	39	39	39	39	39	39	39	39
1947	39	39	39	39	39	39	39	40	40	40	40	40
1948	40	40	40	40	40	40	40	40	40	40	40	40
1949	40	40	40	40	40	40	40	40	41	41	41	41
1950	41	41	41	41	41	41	41	41	41	41	41	41

THE NAMING OF AN UNKOWN PLANET.

The beginner will sometimes notice a bright star-like object which he knows is not a Star, for it is not represented on the maps. He infers that it must be a planet, and he desires to find its name.

It may be assumed that the object must be one of the four bodies—Venus, Mars, Jupiter, or Saturn. With the aid of the Planetary Index it is a simple matter to determine which of the four the unknown object must be.

We will illustrate the process by two examples :—

Example 1.—In May, 1905, a planet is seen in the south at midnight. It is certainly, therefore, not Venus. Plate 43 shows that it is in Scorpio, and its index number is 43. The index numbers in May, 1905, for Mars, Jupiter, and Saturn are respectively, 43, 49, 46. The planet is therefore Mars. This might also be inferred from the Table of Planetory Phenomena. Since the planet is near the meridian at midnight it is close to opposition ; and in 1905 the table shows that Jupiter is in opposition in November, Saturn in August, and Mars in May.

Example 2.—In April, 1916, two planets will be seen at 8 p.m., not far from one another, fairly well up in the west, and a third is in the south. Name the planets. Chart 40 gives the aspect of the sky at 8 p.m. in April, and the planets in the west will be in Taurus or Gemini ; that in the south will be in Leo, and its index number will be 40, the number of the chart. To find the index numbers of the others we must turn to the chart which shows the region between Taurus and Gemini on the central meridian, that is, to Chart 50. The index numbers of the two planets in the west will probably both be 50, but may possibly be 49 or 39. The indexes for April, 1916, give for Venus 50, for Mars 40, for Jupiter 48, and for Saturn 50. The planet in the south is therefore Mars. In the west are Saturn and Venus ; the latter will be the brighter of the two.

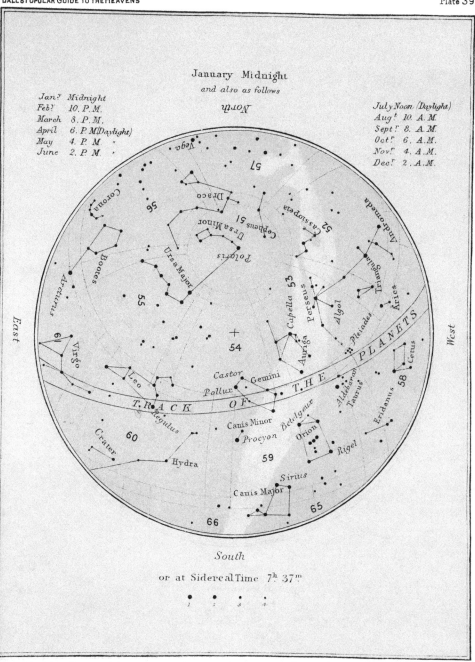

January Midnight
and also as follows

North

Jan.ʸ Midnight
Feb.ʸ 10. P. M.
March 8. P. M.
April 6. P. M.(Daylight)
May 4. P. M. "
June 2. P. M. "

July Noon (Daylight)
Aug.ᵗ 10. A. M.
Sept.ʳ 8. A. M.
Oct.ʳ 6. A. M.
Nov.ʳ 4. A. M.
Dec.ʳ 2. A. M.

East

West

TRACK OF THE PLANETS

South

or at Sidereal Time 7ʰ 37ᵐ

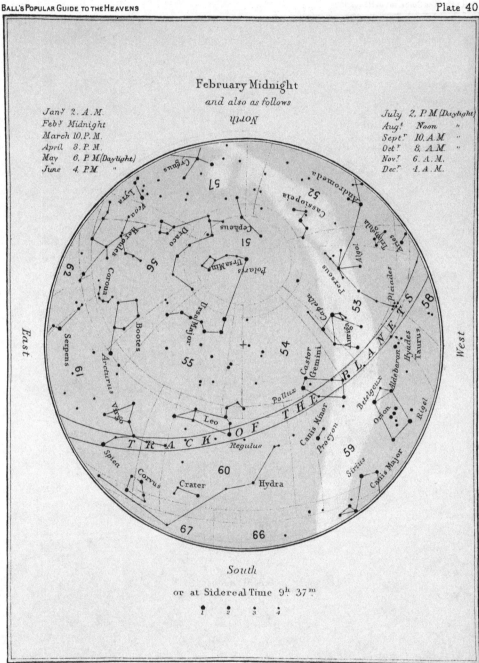

February Midnight
and also as follows

North

Jan.ʸ 2. A.M.
Feb.ʸ Midnight
March 10. P.M.
April 8. P.M.
May 6. P.M. (Daylight)
June 4. P.M. "

July 2. P.M. (Daylight)
Aug.ᵗ Noon "
Sept.ʳ 10. A.M. "
Oct.ʳ 8. A.M. "
Nov.ʳ 6. A.M.
Dec.ʳ 4. A.M.

East

West

South

or at Sidereal Time 9ʰ 37ᵐ

George Philip & Son. Lᵗᵈ

The London Geographical Institute.

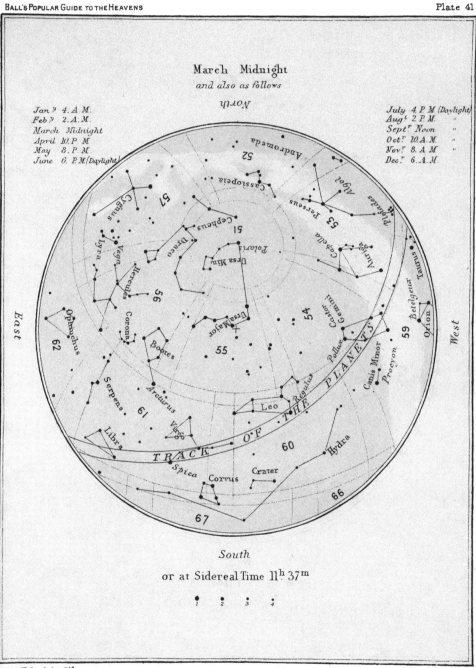

March Midnight
and also as follows

North

Jany 4. A. M.
Feby 2. A. M.
March Midnight
April 10. P. M.
May 8. P. M.
June 6. P. M. (Daylight)

July 4. P. M. (Daylight)
Augt 2. P. M.   "
Septr Noon   "
Octr 10. A. M.   "
Novr 8. A. M.   "
Decr 6. A. M.   "

East

West

South

or at Sidereal Time 11^h. 37^m

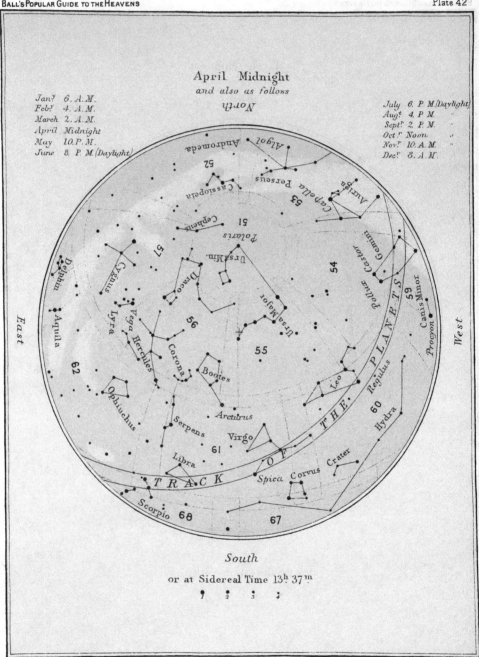

April Midnight
and also as follows

North

Jan.? 6. A.M.
Feb.? 4. A.M.
March 2. A.M.
April Midnight
May 10. P.M.
June 8. P.M. (Daylight)

July 6. P.M. (Daylight)
Aug.? 4. P.M. "
Sept.? 2. P.M. "
Oct.? Noon "
Nov.? 10. A.M. "
Dec.? 8. A.M.

South
or at Sidereal Time 13ʰ 37ᵐ

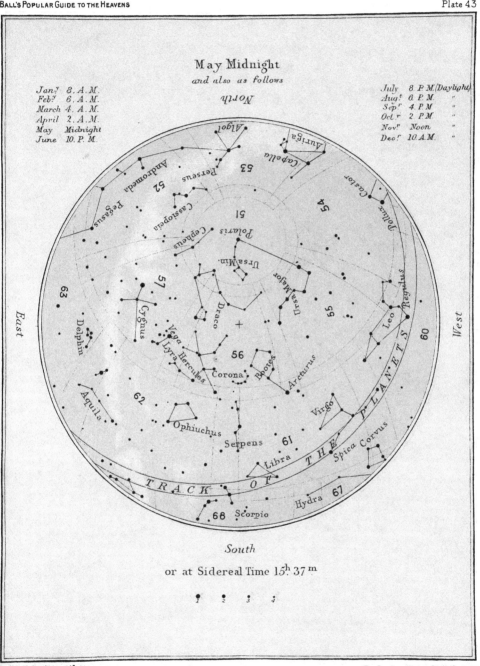

May Midnight
and also as follows

North

Jan.ʸ 8. A.M.
Feb.ᵧ 6. A.M.
March 4. A.M.
April 2. A.M.
May Midnight
June 10. P.M.

July 8. P.M.(Daylight)
Aug.ᵗ 6. P.M. "
Sep.ʳ 4. P.M. "
Oct.ʳ 2. P.M. "
Nov.ʳ Noon "
Dec.ʳ 10.A.M. "

East

West

South
or at Sidereal Time 15ʰ 37ᵐ

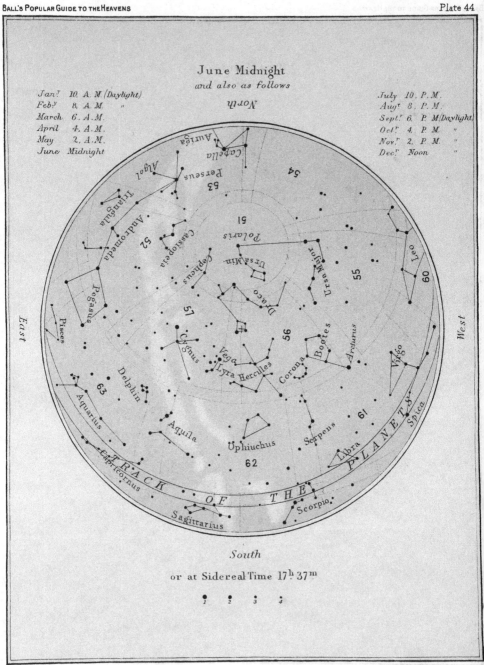

June Midnight
and also as follows

North

Jan.ʸ 10. A.M. (Daylight)
Feb.ʸ 8. A.M.   "
March 6. A.M.
April 4. A.M.
May 2. A.M.
June Midnight

July 10. P.M.
Aug.ᵗ 8. P.M.
Sept.ʳ 6. P.M. (Daylight)
Oct.ʳ 4. P.M.   "
Nov.ʳ 2. P.M.   "
Dec.ʳ Noon   "

Algol
Triangula
Andromeda
Perseus
Capella
Auriga
Cassiopeia
54
53
52
Pegasus
Cepheus
51
Polaris
Ursa Min.
Ursa Major
55
Leo
60
Pisces
57
Draco
56
Bootes
Arcturus
Virgo
Cygnus
Vega
Lyra Hercules
Corona
Delphin
63
Aquila
Ophiuchus
62
Serpens
61
Libra
Spica
Aquarius
Capricornus
TRACK OF THE PLANETS
Sagittarius
Scorpio

South
or at Sidereal Time 17ʰ 37ᵐ

1   2   3   4

George Philip & Son. Lᵗᵈ   The London Geographical Institute

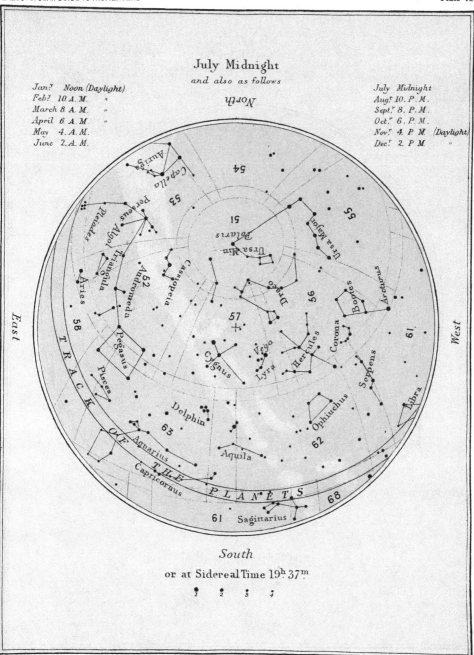

July Midnight
and also as follows

North

George Philip & Son, Lᵗᵃ

The London Geographical Institute.

Plate 46

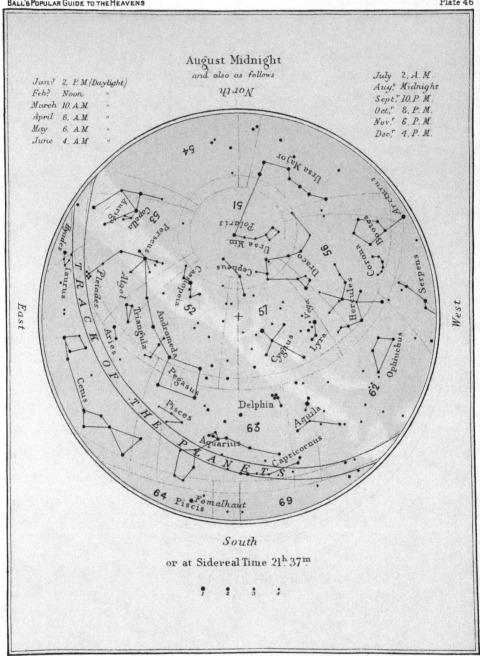

August Midnight
and also as follows

North

Jan.ʸ 2. P. M. (Daylight)
Feb.ʸ Noon   ,,
March 10. A.M.   ,,
April 8. A.M.   ,,
May 6. A.M.   ,,
June 4. A.M   ,,

July 2. A.M.
Aug.ᵗ Midnight
Sept.ʳ 10. P. M.
Oct.ʳ 8. P. M.
Nov.ʳ 6. P. M.
Dec.ʳ 4. P. M.

East

West

South

or at Sidereal Time 21ʰ. 37ᵐ.

George Philip & Son. Lᵗᵈ

The London Geographical Institute.

September Midnight
and also as follows

North

Jan? 4. P.M.(Daylight)	July 4 A.M.(Daylight)
Feb? 2. P.M. "	Aug? 2. A.M.
March Noon "	Sept? Midnight
April 10. A.M. "	Oct? 10. P.M.
May 8. A.M. "	Nov? 8. P.M.
June 6. A.M. "	Dec? 6. P.M.

55

Ursa Major

Boötes

54

Castor

Pollux

Gemini

56

Corona

51

Ursa Minor

Draco

Polaris

Cepheus

57

Vega

Herculis

Ophiuchus

East

59

Betelgeux

Orion

Auriga

Capella

53

Perseus

Cass.

Casseia

Lyra

Cygnus

62

West

TRACK

Hyades

Taurus

Pleiades

Algol

Triangula

Aries

52

Andromeda

Delphin

Aquila

OF

Eridanus

Cetus

58

Pegasus

Pisces

63

Aquarius

Capricornus

THE

PLANETS

69

64   Fomalhaut   Piscis

South
or at Sidereal Time 23ʰ 37ᵐ

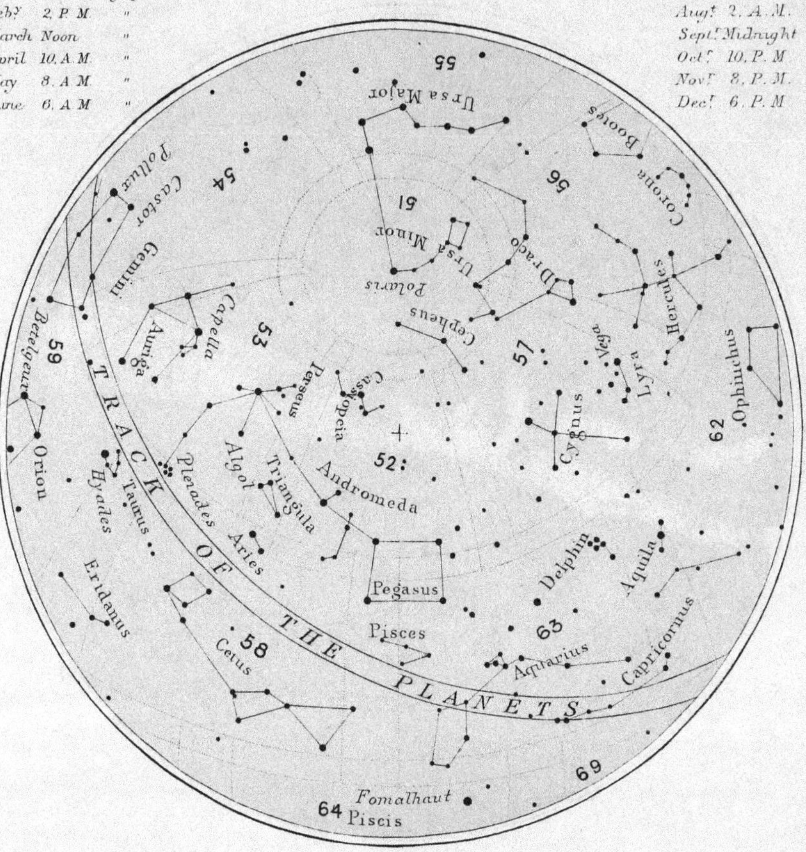

1   2   3   4

George Philip & Son, L.ᵈ   The London Geographical Institute.

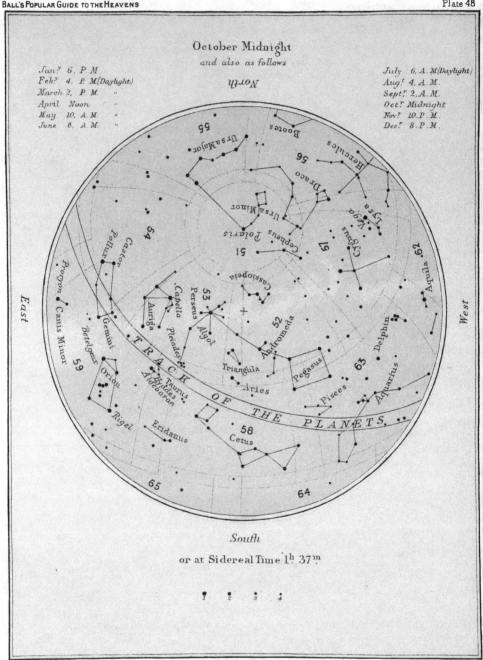

October Midnight
and also as follows

North

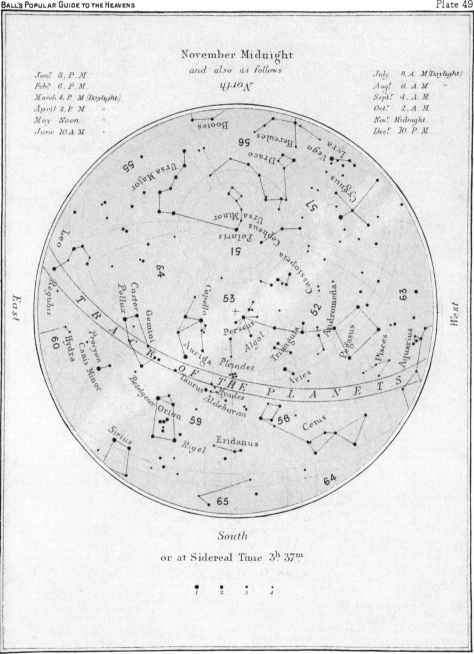

November Midnight
and also as follows

North

Jan.ʸ 8. P. M.
Feb.ʸ 6. P. M.
March 4. P. M (Daylight)
April 2. P. M
May Noon   ,,
June 10. A. M.   ,,

July 8. A. M (Daylight)
Aug.ᵗ 6. A. M   ,,
Sept.ʳ 4. A. M.
Oct.ʳ 2. A. M.
Nov.ʳ Midnight
Dec.ʳ 10. P. M.

East

West

South

or at Sidereal Time 3ʰ 37ᵐ.

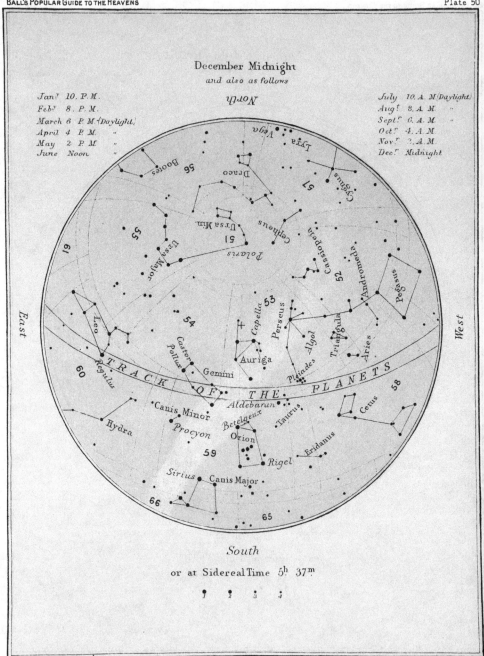

CHAPTER VII.—THE STAR MAPS.

PLATES 51 TO 70.

The student who has made himself familiar with the appearance and movements of the Constellations, and has acquired a facility in identifying the brighter Stars, will soon feel the need of something further. More especially will this be the case if he has the use of a telescope of even moderate dimensions; and it is to meet these requirements that the Star Maps on Plates 51 to 70 have been prepared.

The first step in drawing a map is to decide on the nature of the projection to be employed. It must be understood that no flat maps can give a perfectly faithful representation of a curved surface, and whatever method of projection is resorted to, the result must represent the surface in a more or less distorted form. The Stars appear to be situated on the surface of a sphere, and however we may attempt to depict them, we cannot include any large portion of the sphere exactly as it appears to the eye. The form of projection which I have used in these maps is that known as the conical projection, and in adopting it I follow Argelander, who employed this method in his great Durchmusterung Atlas, which represents more than 300,000 Stars in the Northern Hemisphere alone.

Imagine two cones touching the sphere around the circle of 45° declination, north and south. These are intersected by tangent planes at the Poles, and by a cylinder touching the sphere around the Equator; see adjoining figure. Each star on the sphere is joined to the centre, and the joining line when produced necessarily cuts some one of the enveloping surfaces in a point which is the projection of the star. The equatorial girdle and the two cones are each divided into six equal parts, which admit of being laid out flat; and the eighteen parts thus obtained, together with the two polar planes, make up the twenty maps which represent the entire sphere.

The top and bottom margins of each of these maps, with the exception of the first and last, are divided so as to read Right Ascensions.

Only the hour lines have been drawn on the maps, so as to avoid overcrowding, and for the same reason only the circles corresponding to every tenth degree in Declination have been given. But by the aid of the divisions around the margin, it is easy to read the position of a star, or to enter any desired object with all requisite accuracy. For this purpose it will be found convenient to copy the scale in Declination, which is given on the margin of each map, on the edge of a sheet of paper. If, then, it is desired to enter on the map the position of any object (say a comet) whose R.A. and Declination are known, it is only necessary to set this sheet of paper so that the graduated edge cuts the top and bottom circles at the R.A. of the object, and to put a dot on the map at the point of the scale corresponding to its Declination. In the same way the position of any object entered on the map may be read off. In the case of maps 51 and 70, the method of reading off positions is somewhat different. In these the declination scale will be found on the radius corresponding to 0ʰ., 6ʰ., 12ʰ. or 24ʰ. This scale should be rotated around the centre until it passes through the star whose position is required. The R.A. will then be found at the point of the circumference where the scale cuts it, and the Declination will be read from the scale itself. The epoch for which the places are given is 1880.

It has been arranged that each zone of maps overlaps those north and south of it for a distance of five degrees in Declination, and each map of a zone overlaps those preceding and following it for a space of 40 mins. in R.A.

In order to avoid breaking up conspicuous star groups, I have made the zero, from which the hour circles are measured, pass through the centre of the first map (No. 52) of the intermediate zones, while the same circle divides the first and last maps of the equatorial zone. By this mode of dividing the heavens it has been found possible to comprise each of the more striking configurations of stars within a single map. The only exception is the great square of Pegasus, which will be found on Plates 52, 58, and 63. For convenience in passing from one map to another, the numbers of the plates which represent adjacent portions of the sky have been printed just inside the margin.

In the construction of these maps I have followed, to a great extent, the *Uranométrie Générale* of Houzeau. It contains all the stars visible to the naked eye under the most favourable circumstances, and the number amounts to nearly 6,000.

In the nomenclature of the stars, however, I have departed considerably from Houzeau, doing away in general with letters (other than those of the Greek alphabet), and substituting, wherever possible, Flamsteed's numbers.

I have followed Houzeau throughout in the estimation of star magnitudes, as by so doing I obtained a uniform system over the whole sky, both in the Northern and Southern Hemispheres, determined by a single observer in the same climate and within a short time. His work, besides, is more recent than that of Argelander, Heis, or Behrmann. I have further, for simplicity, limited the number of magnitudes given by Houzeau to six, namely, 1, 2, 3, 4, 5, and 6, which will be found sufficient for all ordinary purposes. These I have indicated on the maps themselves, as shown by the scale at the foot of each map, where, in addition to the size of the dot representing the star, its magnitude is denoted by the number of rays diverging from it. Thus all stars of the first magnitude possess 6 rays, those of the second magnitude 5 rays, and so on. The magnitude of a star may be found by subtracting the number of the rays from seven, except for the sixth magnitude, in which case the single ray has been omitted, stars of this class being represented by a simple dot.

Throughout the maps a large number of the stars will be found accompanied by the letter D. This signifies that the star, though appearing as a single body to the naked eye, is in reality double. This does not however denote necessarily any physical connection between the two stars, but merely that the point of light thus characterized will be found to break up into two with the aid of a small telescope. In many cases there is a real connection; the two stars form a binary pair, and revolve round their common centre of gravity under the action of their mutual gravitation; a number of the most interesting cases will be found in the list of select telescopic objects. In other cases, the connection is merely apparent; the stars happen to be nearly in the same direction as seen from the solar system; but they are probably at very different distances from us.

VARIABLE STARS.

A number of stars will be found in the maps marked (Var.), which indicates that their light is not constant, but is subject to fluctuations, in some cases perfectly regular, so that their times of greatest brightness can be predicted to the minute ; in other cases, more or less regular, so that their maxima can be predicted to within a week or a month ; in yet other cases, so irregular that no law has yet been found for their variation. In the following tables I have given those variable stars which are plainly visible to the naked eye at least at their maximum brightness. They may be easily identified from the star maps. There are many hundreds of variable stars which never rise to naked eye brilliancy, but for their identification means more elaborate than those of the present work are required. Without detailed charts of the fainter stars it is very hard to find them. For information regarding them the student is referred to the lists published in the *Companion to the Observatory;* for their identification one needs the star maps of Argelander's Durchmusterung, and of the Cordoba Durchmusterung, or, better still, the Atlas of Variable Stars, published by Father Hagen, of Georgetown, D.C.

"ALGOL" VARIABLES.

There is a small class of stars whose type of variation is characteristic, and of which the explanation is certain. The stars are partially eclipsed by a dark companion revolving round them. Algol is the typical star, which has given its name to the group.

The type of variation of a variable star is best shown by the light curve. The light curve of an invariable star is a horizontal straight line. A decrease of magnitude is shown by the line dipping down, and if the curve is carefully drawn, so that equal distances horizontally denote equal intervals of time, while equal intervals vertically denote equal changes of magnitude, the curve is a complete representation of the light variability.

Thus, for Algol, the light curve is drawn thus :—

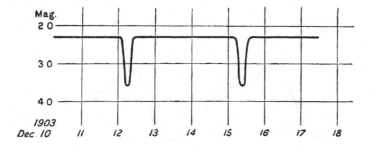

and the interpretation is as follows. During the greater part of the time the light of Algol is uniformly of magnitude, 2·3 ; but every 2 d. 20 h. 49 m. the star drops down to magnitude 3·6, and without pausing regains its original light, the whole change occupying 9 h. 20 m. The figure further shows minima are predicted for 1903 (December 12th, at about 9 o'clock in the evening, and for December 15th, at 6 o'clock).

The three Algol variables which are visible to the naked eye are :—

Name.	R.A.	Decl.	Range of Variation.	Period.
β Persei (Algol)..............	3ʰ. 0ᵐ.	+ 40° 29′	2·3 to 3·6	2ᵈ. 20ʰ. 49ᵐ
λ Tauri	3 54	+ 12 9	3·4 to 4·2	3 22 52
δ Libræ	14 55	− 8 2	5·0 to 6.2	2 7 51

These may be readily found from the star maps ; and the times when they undergo eclipse may be found in the *Companion to the Observatory*.

SHORT PERIOD VARIABLES.

Our second class comprises stars which are regularly variable in periods of not many days, but whose variation is not due to eclipse by a dark companion. In many cases the spectroscope has shown that these stars are binary, and it is probably so in all ; but the way in which the duplicity of the star explains its variation is quite uncertain. The light variation is continuous, and by the shape of the light curves, these short period variables are divided into two classes. In the first, the rise to maximum is steep and the fall to minimum gentle ; in the second the curve is symmetrical. The two types may be represented thus :—

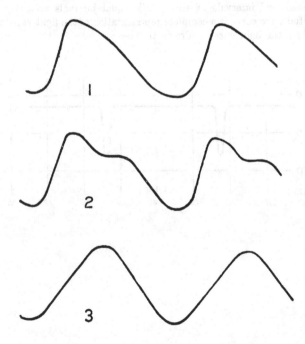

The following is a list of short period variables which rise above naked eye brightness. They may be found upon the maps whose numbers are given in the last column.

	R.A.		Decl.		Range.	Period in Days.	Map.
	h. m.				m. m.		
ζ Geminorum	6	57	+ 20°	45′	3·7 to 4·5	10·15	59
V Puppis	7	55	− 48	5	4·4 to 5·2	2·25	66
N Velorum	9	28	− 56	30	3·4 to 4·4	Not well known	66
l Carinæ	9	42	− 61	57	3·7 to 5·2	35·05	66
X 3 Sagittarii	17	40	− 27	47	4 to 6	7·01	68
W γ₁ Sagittarii	17	57	− 29	35	4·8 to 5·8	7·59	68
Y Sagittarii	18	14	− 18	55	5·8 to 6·6	5·77	62
κ Pavonis.................	18	45	− 67	23	4·0 to 5·5	9·10	69
β Lyræ	18	46	+ 33	13	3·4 to 4·5	12·91	57
R Lyræ	18	52	+ 43	48	4·0 to 4·7	46·4	57
η Aquilæ	19	46	+ 0	42	3·5 to 4·7	7·18	62
S Sagittæ	19	51	+ 16	19	5·6 to 6·4	8·38	62
T Vulpeculæ..............	20	46	+ 27	48	5·5 to 6·5	4·44	57
δ Cephei	22	55	+ 57	48	3·7 to 4·9	5.37	52

In case the limits of variation are not well determined, the magnitudes are given in round numbers.

LONG PERIOD VARIABLES.

These stars have periods which average between 300 and 400 days, and only one regular variable is known to have a period greater than 600 days. Almost all these stars show at maximum bright lines of hydrogen in their spectra ; they are mostly capricious in their behaviour, rising higher at some maxima than at others, and sinking lower at some minima. Nothing is certainly known as to the cause of their variability, but there is no reason to suppose that it is due in any way to a companion.

The more conspicuous of the long period variables are as follows :—

		R.A.		Decl.		Range.	Period in Days.	Map.
		h. m.						
o	Ceti (Mira)	2	13	− 3°	32′	3 : to 9 :	331	58
η	Geminorum	6	8	+ 22	32	3·2 to 4 :	231	59
L₂	Puppis	7	10	− 44	27	3·5 to 6·3	140	66
R	Carinæ	9	29	− 62	15	5 : to 10 :	309	66
R	Hydræ	13	23	− 22	40	4 : to 9·7	425	61
χ	Cygni..............	19	46	+ 32	37	5 : to 13·5	406	57
R	Cassiopeiæ	23	52	+ 50	43	5 : to 12	429	52

Where the magnitude given is followed by a colon (:) it is subject to irregularity.

IRREGULARLY VARIABLE STARS.

The stars in the first three classes vary in periods which are, on the whole, the less regularly followed the longer the period. We now come to a class of stars which fluctuate so irregularly that no law of variation has yet been discovered. In such cases the magnitudes given are generally the highest and lowest which have been observed, and there is every probability that the range may be exceeded at some time or another.

		R.A. h. m.	Decl.	Range.	Map.
T	Ceti	0 16	−20° 44′	5 to 7	58
α	Cassiopeiæ	0 34	+55 53	2·2 to 2·8	52
ρ	Persei......................	2 57	+38 22	3·4 to 4·2	53
ε	Aurigæ	4 53	+43 38	3·0 to 4·5	53
α	Orionis	5 49	+ 7 23	0·5 to 1·4	59
U	Hydræ	10 32	−12 46	4·5 to 6	60
η	Carinæ (η Argus).........	10 40	−59 3	>1 to 7·4	67
W	Bootis......................	14 38	+27 2	5·2 to 6.1	56
R	Coronæ	15 44	+28 32	5·8 to 13·0	56
g	Herculis....................	16 25	+42 8	5 to 6	56
α	Herculis	17 9	+14 32	3·1 to 3·9	62
u	Herculis	17 13	+33 14	4·6 to 5·4	56
R	Scuti	18 41	− 5 50	5 to 9	62

TEMPORARY OR "NEW" STARS.

So called "new" stars are stars that have suddenly appeared once, and then faded away They all seem to follow essentially the same course. They blaze up very suddenly, gradually fade away, often with fluctuations, then appear to turn into small gaseous nebulæ, and finally, as has recently been shown, become merely very faint stars, very probably their original condition. There is no theory that satisfactorily accounts for more than a part of the facts known about them.

The following is a list of temporary stars which have been observed, arranged in order of discovery.

		Greatest br.	R.A. h. m. 1880.	Decl.	Map.
1572	Tycho's Nova in Cassiopeia	>1	0 18	+ 63° 29	52
1604	Nova in Serpentarius, discovered by Fabricius ...	>1	17 23	− 21 22	62
1670	Nova in Vulpecula, discovered by Anthelm	3·0	19 43	+ 27 1	57
1848	Nova Ophiuchi, discovered by Hind	5·5	16 53	− 12 43	62
1860	Nova Scorpii, discovered by Auwers in the cluster Messier 80.................................	7·0	16 10	− 22 41	62
1866	Nova Coronæ, discovered by Birmingham	2·0	15 55	+ 26 15	56
1876	Nova Cygni, discovered by Schmidt	3·2	21 37	+ 42 18	57
1885	Nova Andromedæ, discovered by Hartwig in the Andromeda Nebula	7·0	0 37	+ 40 37	52
1887	Nova Persei, No. 1., discovered on Harvard photographs by Mrs. Fleming, 1895	9·0	1 54	+56 9	53
1891	Nova Aurigæ, discovered by Anderson, January, 1892, and afterwards found on a photograph taken at Harvard, December 10th, 1891.........	4·5	5 24	+ 30 21	53
1893	Nova Normæ, discovered by Mrs. Fleming on Harvard photographs	7	15 20	− 50 9	68
1895	Nova Carinæ, discovered by Mrs. Fleming on Harvard photographs	8	11 3	− 61 18	67
1895	Nova Centauri, discovered by Mrs. Fleming on Harvard photographs	7·2	13 33	− 31 2	67
1898	Nova Sagittarii, discovered by Mrs. Fleming, 1899	4·7	18 55	− 13 19	62
1899	Nova Aquilæ, discovered by Mrs. Fleming on Harvard photographs, 1900	7			
1901	Nova Persei, discovered by Anderson..............	0·0	3 23	+ 43 30	53
1903	Nova Geminorum, discovered by Turner, on Oxford Astrographic Chart plates	7	6 36	+ 30 4	54

METEOR SHOWERS.

In some of the maps will be found an asterisk, closely accompanied by a date. This marks a region from which meteors may be expected about the date in question—in the language of the meteor observer, it is the radiant point of a meteor shower. It may be well, however, to caution the beginner against expecting too much of a display from such a shower. For most of the points a dozen meteors in the night rank as a rich display, and in many years the radiants are almost completely quiescent. Many of the radiants seem to be persistent, furnishing occasional meteors throughout the year : a fact for which no easy explanation is forthcoming ; for if the meteors of one radiant really belonged to one system, one would expect that radiant to shift during the year, as the radiant of the Perseids is supposed to do. For information as to the smaller and sparser showers reference may be made to the memoir by Mr. Denning (Memoirs, Royal Astron. Soc., Vol. LIII.). A list of radiant points is given in the *Companion to the Observatory*. We may limit ourselves here to four principal showers.

The LYRIDS. April 19-22. R.A. 18h. 8m., Declination + 34°

In some years this radiant gives a number of meteors ; in others they are almost entirely absent.

The PERSEIDS. August 9-11. R.A. 3h. 0m., Declination + 57°

These are the well-known August meteors, which are, on the whole, the most reliable shower of the year. They leave long, yellow streaks. For a month before the date of maximum, similar meteors appear from radiants lying to the west of the above place, and it is believed that they are a real case of a moving radiant.

The LEONIDS. November 12-14. R.A. 10h. 0m., Declination + 23°

This shower is visible more or less every November, giving swift meteors with greenish streaks. For a number of centuries they had given brilliant displays every 33 years, but in 1899 and 1900 the expected display completely failed. It is fairly certain that the swarm had been perturbed by the planet Jupiter. There is reason to believe that the swarm was captured and introduced into the solar system by the planet Uranus in the year 126 A.D.

The ANDROMEDIDS. November. R.A. 1h. 40m., Declination + 43.

This is the shower associated with the lost Biela's Comet. Rich displays may be expected with some certainty every thirteen years. Of late years the perturbations of the planet upon the swarm have had the effect of throwing back the date of maximum and number of days, and it is impossible at present to give exact dates for the future.

PLATES 71 AND 72.

As I have already pointed out, the region of the sky which corresponds to any one of the general series of maps, is indicated by the dotted lines in the series of monthly maps (Plates 39 to 50). This, however, is chiefly useful at localities about the latitude of the British Islands. For the convenience of those living in other latitudes, to whom it is hoped this Atlas will recommend itself, as well as to enable the student at home to choose the maps suitable for his purpose with greater rapidity, I have added the Northern and Southern Index Maps (Plates 71 and 72). In these the principal constellations are marked, and the outlines of each map of the general series, with the numbers of the corresponding plates in bold figures. Each Index Map includes from the Pole to 25° beyond the Equator, so that both contain the series of Equatorial maps. Around the circumferences is marked each hour of R.A. The Declination is not indicated, but it can be ascertained with sufficient accuracy for the purpose of finding the required map by remembering that the Equatorial zone extends to 25° Declination, and the intermediate zones to 70° Declination, while each zone overlaps that above and below it by 5°.

PRECESSION.

The Precession of the Equinoxes, or the slow motion of the Earth's axis, in consequence of which the intersection of the Equator with the Ecliptic travels along the latter, brings about a constant change in the R.A. and Declination of the Stars from year to year. It is thus clear that the values of these quantities as read from the maps will only be strictly accurate at the epoch for which the maps are drawn. In order to find the R.A. and Declination for any other date, it is necessary to apply a correction for this precessional effect, and if it is desired to mark upon the maps the position of any star or other object whose co-ordinates are given for a date different from that of the Atlas, a similar correction must be applied.

It must, however, be borne in mind that no change takes place from this cause in the relative position of the stars,—the effect being merely to give the whole system of Right Ascension and Declination circles a shift, and thus to alter the positions of all the stars with regard to them.

For accurate astronomical work, the correction for precession must in general be computed to a small fraction of a second, and elaborate tables have been prepared to facilitate this operation ; but for all purposes coming within the scope of the present work, the following tables will be found amply sufficient.

That given on the next page contains the correction to the R.A. for 10 years' precession. The quantity found in the table is to be added, with the sign there indicated to the R.A. at any time, in order to obtain the R.A. for an epoch 10 years later, or it is to be subtracted to find the R.A. at an epoch 10 years earlier. For intervals other than 10 years a proportional allowance must be made. The top and bottom lines contain the Declination, and the first and last columns the R.A.

For most purposes it will be sufficient in finding the precession to take the R.A. to the nearest whole hour, and the Declination to the nearest multiple of 10 degrees. If the star is situated in the Northern Hemisphere, we find its Declination in the first or last line, and run the eye down the corresponding column till we reach the line which contains the star's R.A. in the first column ; the corresponding figure in the table is the precession in R.A. for 10 years. If the star is in the Southern Hemisphere, we look for its Declination as before, but we find its R.A. in the last column.

The second table, containing the correction to the Declination for 10 years, is still more simple. We have merely to enter it with the nearest hour of R.A. in the extreme columns, and we find in the central column the corresponding correction to the Declination. For all R.A.'s found on the left side the correction is positive, and negative for all those on the right side.

The signs of the precessions given in both tables show the correction necessary to bring the star's place up to a subsequent date ; to bring it back to an earlier date the signs must be altered. The table of precession in R.A. extends to 70° north and south of the Equator, so that it is applicable to all the stars except those around the North and South Poles, contained in Plates 51 and 70.

TABLE FOR PRECESSION IN R.A.

R.A. for N. Decl.	0°	10°	20°	30°	40°	50°	60°	70°	R.A. for S. Decl.
h.　　h.	m.	m.	m.	m.	m.	m.	m.	m.	h.　　h.
18 or 18	+0·51	+0·47	+0·43	+0·38	+0·33	+0·25	+0·13	−0·10	6 or 6
19 „ 17	·51	·47	·43	·39	·33	·26	·14	·08	5 „ 7
20 „ 16	·51	·48	·44	·40	·35	·28	·18	−0·02	4 „ 8
21 „ 15	·51	·48	·45	·42	·38	·32	·24	+0·08	3 „ 9
22 „ 14	·51	·49	·47	·45	·42	·38	·32	·21	2 „ 10
23 „ 13	·51	·50	·49	·48	·46	·44	·41	·35	1 „ 11
0 „ 12	·51	·51	·51	·51	·51	·51	·51	·51	0 „ 12
1 „ 11	·51	·52	·53	·54	·56	·58	·61	·67	23 „ 13
2 „ 10	·51	·53	·55	·58	·61	·64	·70	·82	22 „ 14
3 „ 9	·51	·54	·57	·60	·64	·70	·78	0·94	21 „ 15
4 „ 8	·51	·55	·58	·62	·67	·74	·85	1·04	20 „ 16
5 „ 7	·51	·55	·59	·64	·69	·77	·88	1·10	19 „ 17
6 „ 6	+0·51	+0·55	+0·59	+0·64	+0·70	+0·78	+0·90	+1·12	18 „ 18
R.A. for N. Decl.	0°	10°	20°	30°	40°	50°	60°	70°	R.A. for S. Decl.

TABLE FOR PRECESSION IN DECLINATION.

R.A.		Precession.	R.A.		
h.	h.	°	h.		h.
0 or	24	+ 0·06 −	12	or	12
1 „	23	·05	13	„	11
2 „	22	·05	14	„	10
3 „	21	·04	15	„	9
4 „	20	·03	16	„	8
5 „	19	·01	17	„	7
6 „	18	·00	18	„	6

Example.—The star Capella is situated in 1880 in R.A. 5 h. 8 m., Declination + 45°·9 : find what its R.A. and Declination will be in 1905.

Entering the first Table with R.A. 5 h., and Declination 50°, we find 10 years' precession in R.A. is + 0·77 m. Hence the corresponding correction for 25 years will be to the nearest whole minute + 2 m.

Entering the second Table with R.A. 5 h., we find 10 years' precession in Declination is + 0°·01, hence to the tenth of a degree the correction for 25 years is negligible, so that we find in 1905 R.A. = 5 h. 8 m. + 2 m. = 5 h. 10 m., and Declination = + 45°·9.

If it were required to find the place of the star at the beginning of the century (*i.e.*, 80 years previously), we have to multiply + 0·77 m. and + 0°·01 by − 8, and we find the corrections − 6 m. and − 0°·1, so that the place of this star in 1800 is R.A. 5 h. 2 m., Declination + 45°·8.

As another example, let us find the R.A. and Declination of ω Draconis in 1940. Its place in 1880 is 17 h. 38 m. ; + 68°·8. We find from the Tables − 0·10 m. as correction for 10 years' precession, and 0°·00 as the correction in Declination ; we thus obtain for 1940 R.A. = 17 h. 38 m. − 0·6 m. = 17 h. 37 m. to the nearest minute, and Declination + 68°·8.

Once more, suppose that in 1950 it is announced that a comet has been seen in R.A. 3 h. 42·9 m., and Declination + 23°·96. We find the precession in R.A. and Declination from the tables to be, for 10 years, + 0·58 m. and + 0°·03. Hence, to bring the place back to 1880, we have the correction — 4·1 m. and — 0°·21. We thus have

	Comet's R.A.		Comet's Declination.
	h. m.		°
1950.............................	3 42·9	...	+ 23·96
Correction for Precession...	— 4·1	...	— 0·21
1880.............................	3 38·8	...	+ 23·75

That is to say, the place occupied by the comet is indicated on these maps by the figures just found for 1880, so that it would be found at the time of the announcement in the centre of the group of the Pleiades.

The star maps of this work were drawn for the Epoch 1880, and as has been already explained, they no longer give the R.A. and Declination of the stars as they are measured to-day. It will be convenient if we add a table showing how much the system of circles should be shifted on each map relatively to the stars, to make them approximately right for the Epoch 1900, to which for a number of years the places of the stars will generally be referred.

The following table gives the number of the map, and the amount the system of circles must be shifted in millimetres.

52.	0·4 right.	0.2 down.	58.	0·4 right.	0.2 down.	64.	0·4 right.	0.2 down.
53.	0·6 ,,	0.1 ,,	59.	0·4 ,,	0.0	65.	0·3 ,,	0.1 ,,
54.	0·6 ,,	0.1 up	60.	0·4 ,,	0.2 up	66.	0·3 ,,	0.1 up
55.	0·4 ,,	0.2 ,,	61.	0·4 ,,	0.2 ,,	67.	0·4 ,,	0.2 ,,
56.	0·3 ,,	0.1 ,,	62.	0·4 ,,	0.0	68.	0·6 ,,	0.1 ,,
57.	0·3 ,,	0.1 down.	63.	0·4 ,,	0.2 down.	69.	0·6 ,,	0.1 ,,

It will be seen that for the purposes of these maps the change is almost insensible.

The London Geographical Institute.

George Philip & Son, Ltd.

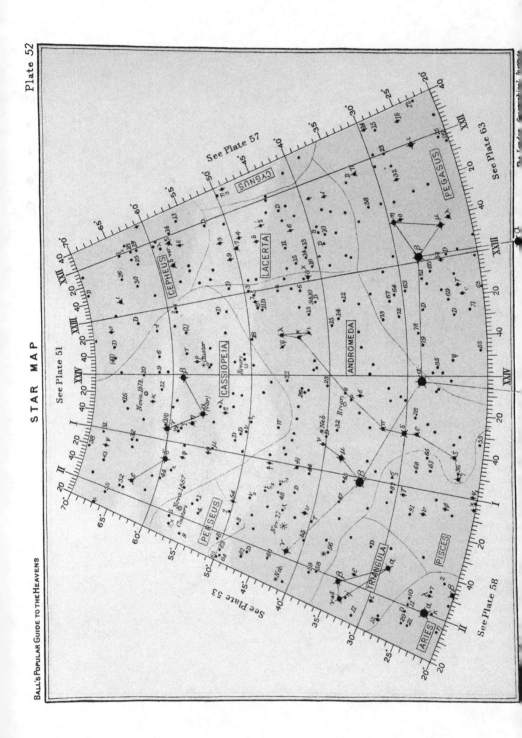

Plate 52

See Plate 51

See Plate 57

See Plate 63

See Plate 53

See Plate 58

CEPHEUS

CYGNUS

LACERTA

CASSIOPEIA

PERSEUS

ANDROMEDA

PEGASUS

TRIANGULA

ARIES

PISCES

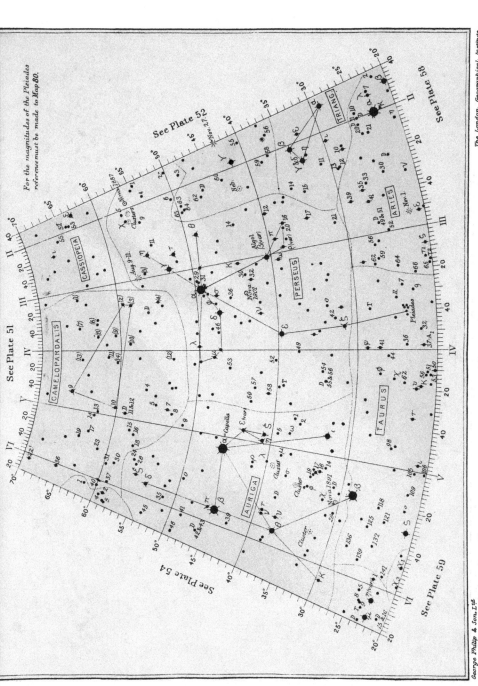

For the magnitudes of the Pleiades reference must be made to Map 80.

See Plate 51

See Plate 52

See Plate 58

See Plate 50

See Plate 54

CASSIOPEIA

CAMELOPARDALIS

AURIGA

TAURUS

PERSEUS

ARIES

TRIANG.

The London Geographical Institute.

George Philip & Son, L.td.

Plate 54

STAR MAP

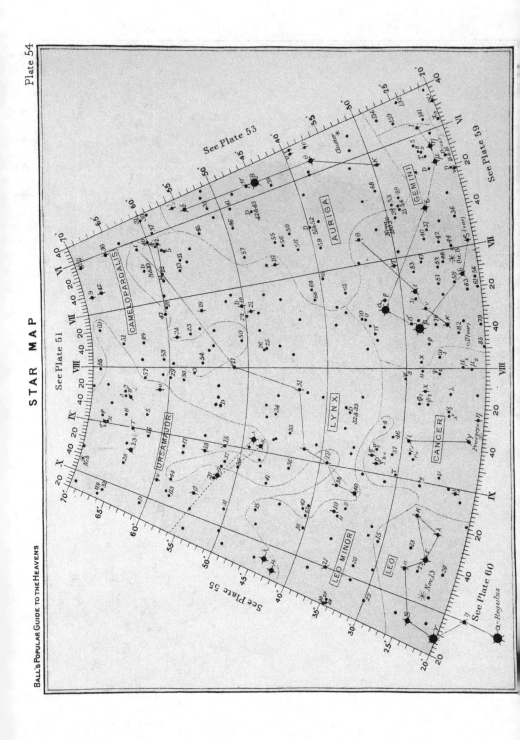

See Plate 53

See Plate 51

See Plate 55

See Plate 59

See Plate 60

CAMELOPARDALIS

URSA MAJOR

LYNX

AURIGA

GEMINI

CANCER

LEO MINOR

LEO

α-Regulus

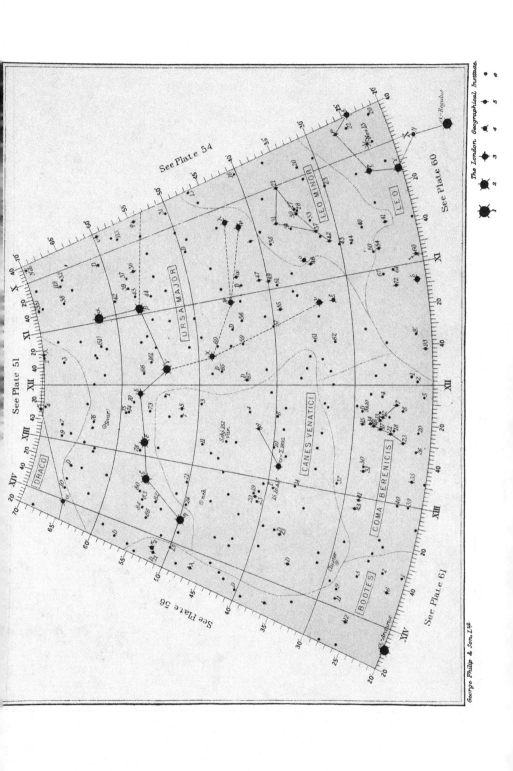

See Plate 54

See Plate 51

See Plate 56

See Plate 60

See Plate 61

URSA MAJOR

LEO MINOR

LEO

DRACO

CANES VENATICI

COMA BERENICIS

BOOTES

The London Geographical Institute.

George Philip & Son, L.ᵗᵈ

Plate 56

STAR MAP

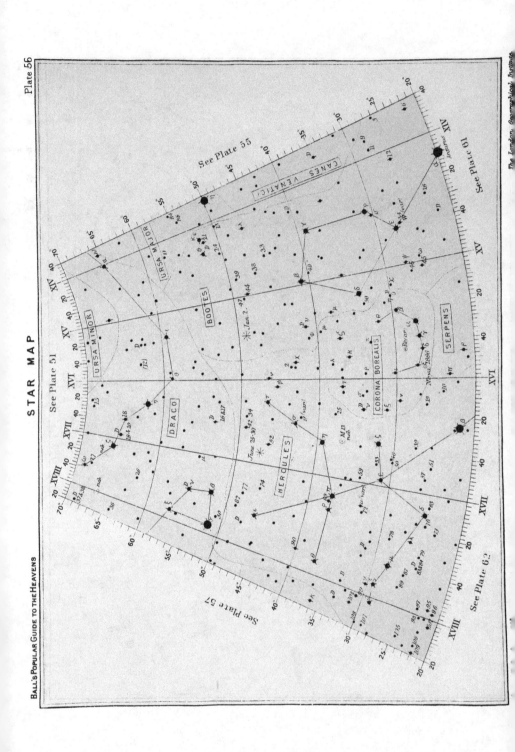

See Plate 55

See Plate 51

See Plate 57

See Plate 61

See Plate 63

URSA MAJOR

URSA MINOR

BOÖTES

DRACO

HERCULES

CORONA BOREALIS

SERPENS

CANES VENATICI

The London Geographical Institute

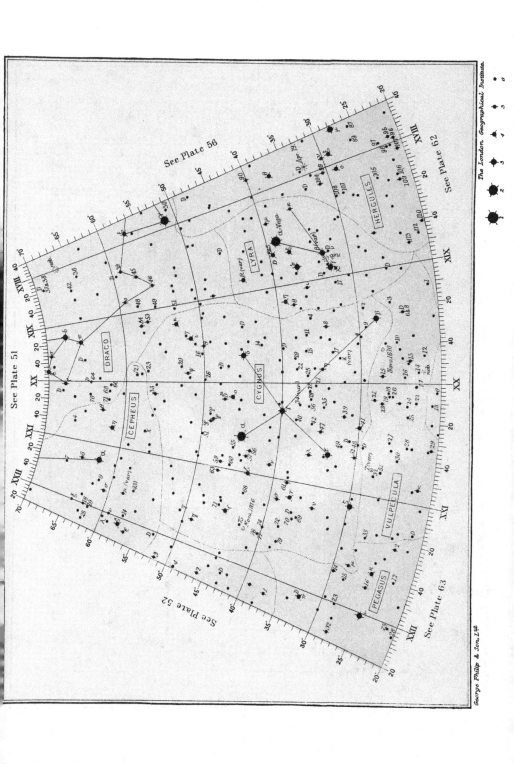

Plate 58

STAR MAP

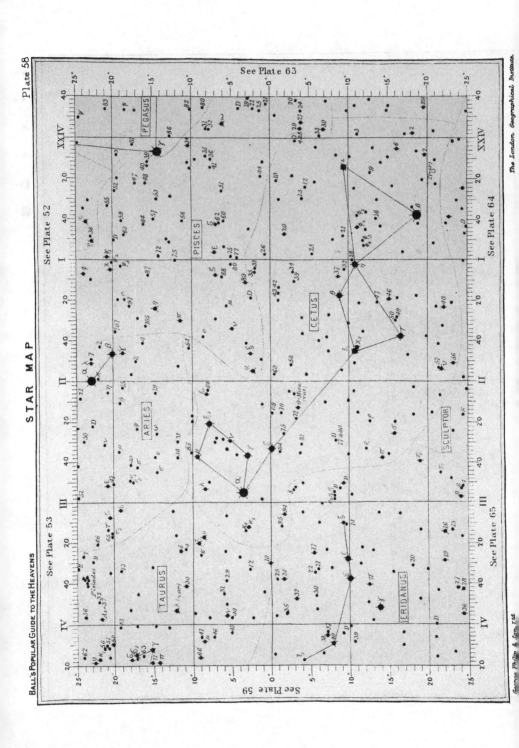

See Plate 63

See Plate 52

See Plate 53

See Plate 59

See Plate 64

See Plate 65

PEGASUS

PISCES

ARIES

TAURUS

CETUS

ERIDANUS

SCULPTOR

The London Geographical Institute.

George Philip & Son, Ltd

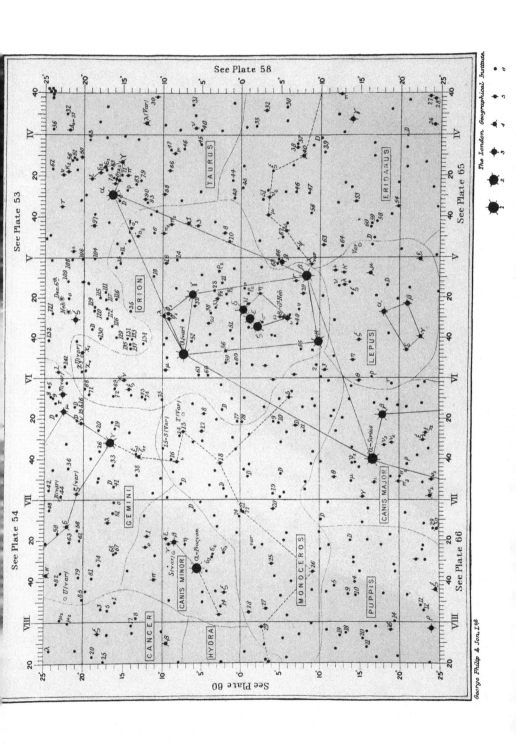

See Plate 53

See Plate 54

See Plate 65

See Plate 66

TAURUS

ERIDANUS

ORION

LEPUS

GEMINI

MONOCEROS

PUPPIS

CANIS MAJOR

CANIS MINOR

CANCER

HYDRA

Plate 60

STAR MAP

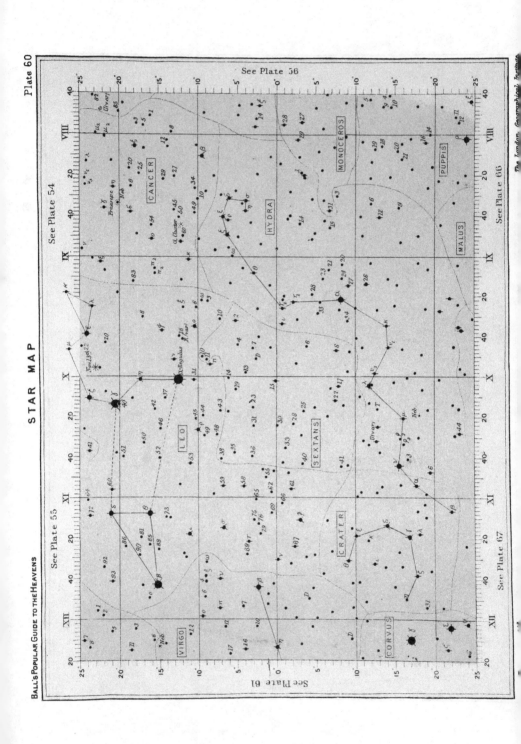

See Plate 56

See Plate 54

See Plate 55

See Plate 61

See Plate 66

See Plate 67

The London Geographical Institute

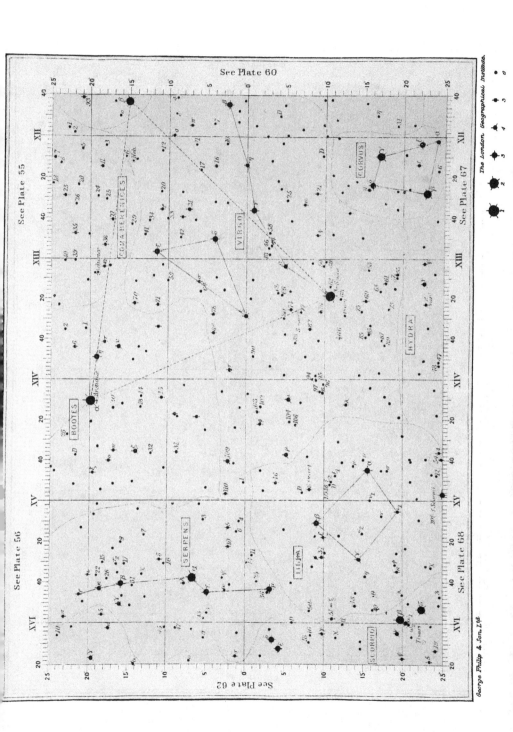

See Plate 55

See Plate 56

See Plate 62

See Plate 67

See Plate 68

XII

XIII

XIV

XV

XVI

COMA BERENICES

BOOTES

SERPENS

VIRGO

CORVUS

HYDRA

LIBRA

SCORPIO

The London Geographical Institute.

George Philip & Son, Ltd.

STAR MAP

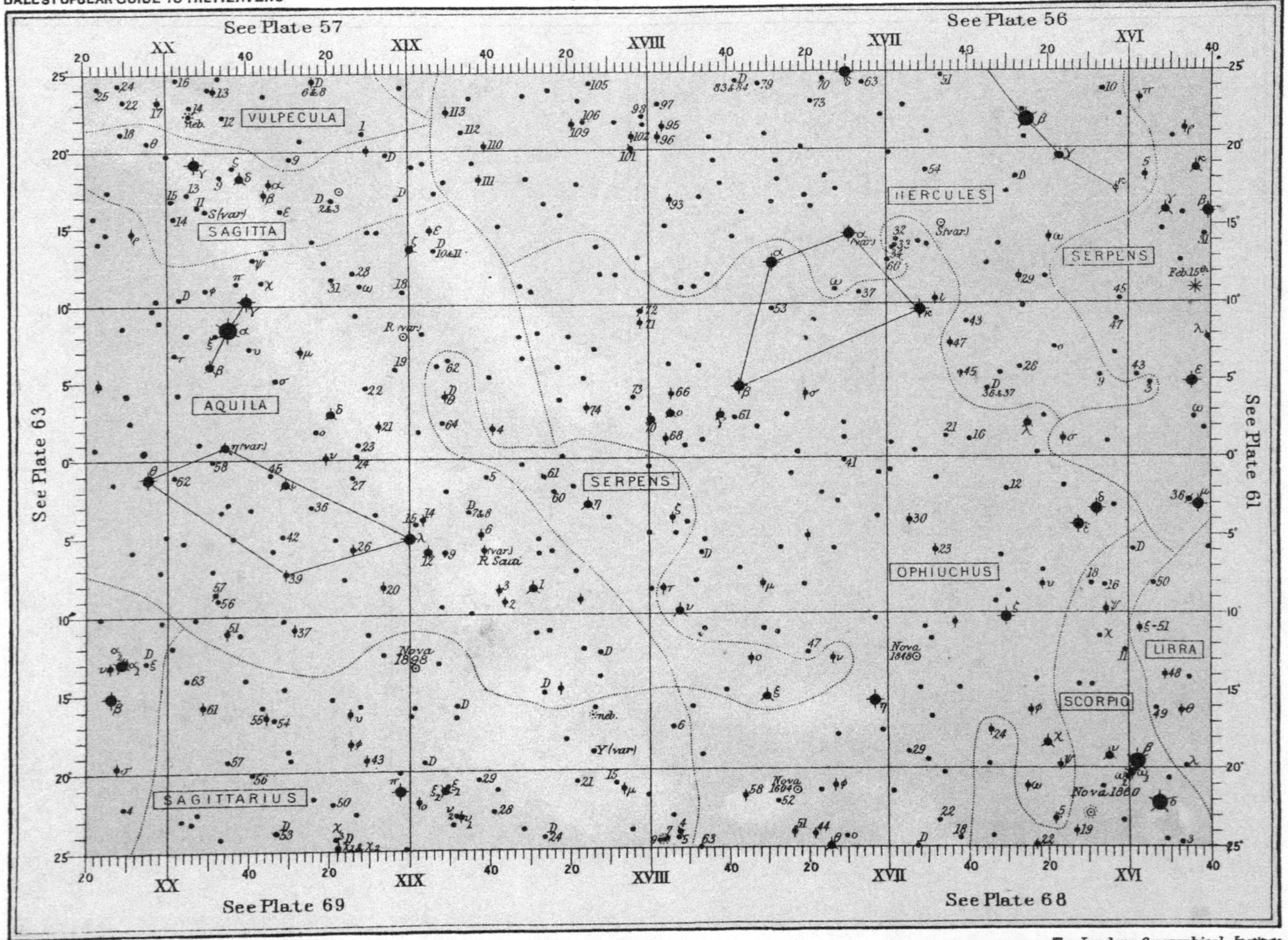

BALL'S POPULAR GUIDE TO THE HEAVENS

Plate 62

See Plate 61

See Plate 56

See Plate 57

See Plate 68

See Plate 69

See Plate 63

The London Geographical Institute.

George Philip & Son, Ltd.

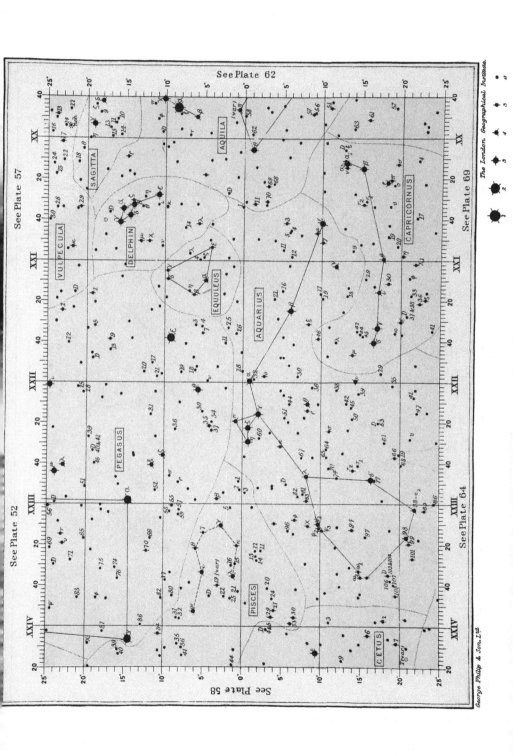

See Plate 69

See Plate 64

AQUILA

SAGITTA

VULPECULA

DELPHIN

EQUULEUS

AQUARIUS

CAPRICORNUS

PEGASUS

PISCES

CETUS

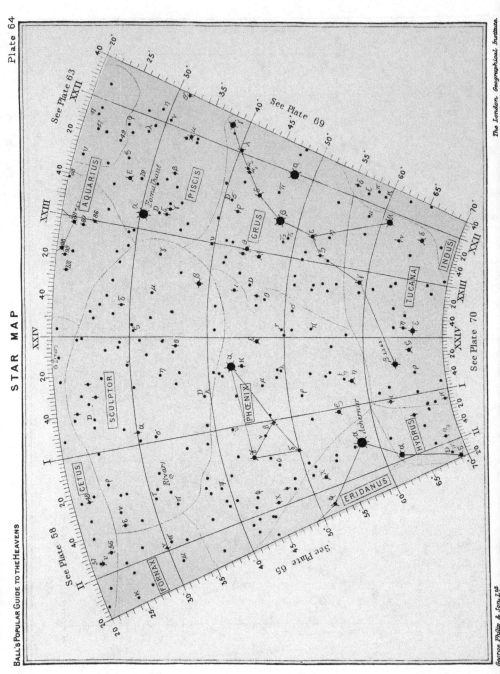

Plate 64

STAR MAP

See Plate 63

See Plate 58

See Plate 69

See Plate 70

See Plate 65

AQUARIUS

PISCIS

GRUS

INDUS

TUCANA

SCULPTOR

PHŒNIX

HYDRUS

CETUS

ERIDANUS

FORNAX

Fomalhaut

Achernar

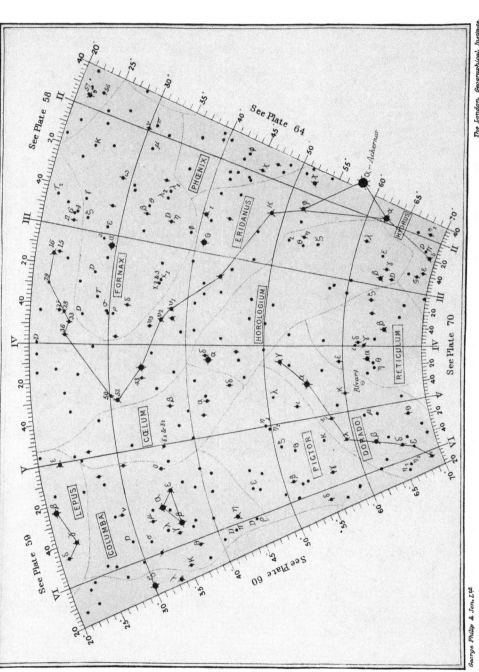

See Plate 58

See Plate 64

See Plate 50

See Plate 60

See Plate 70

The London Geographical Institute.

George Philip & Son, Ltd.

PHŒNIX

ERIDANUS

FORNAX

HOROLOGIUM

HYDRUS

RETICULUM

CŒLUM

DORADO

PICTOR

LEPUS

COLUMBA

α—Achernar

Plate 66

STAR MAP

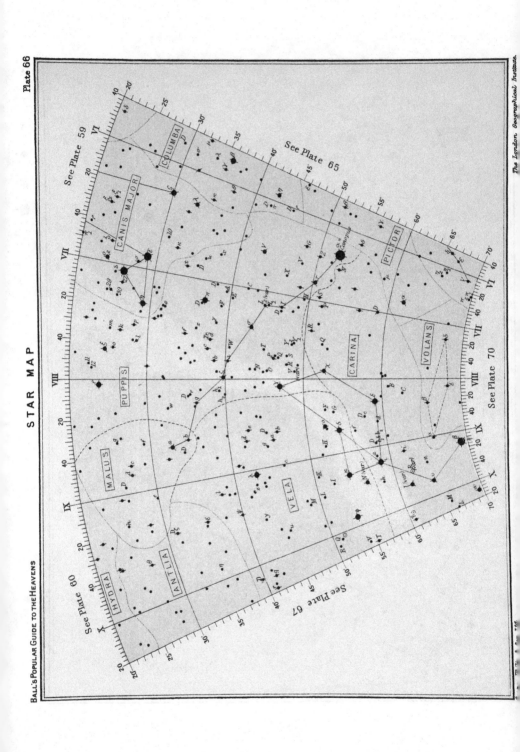

The London Geographical Institute.

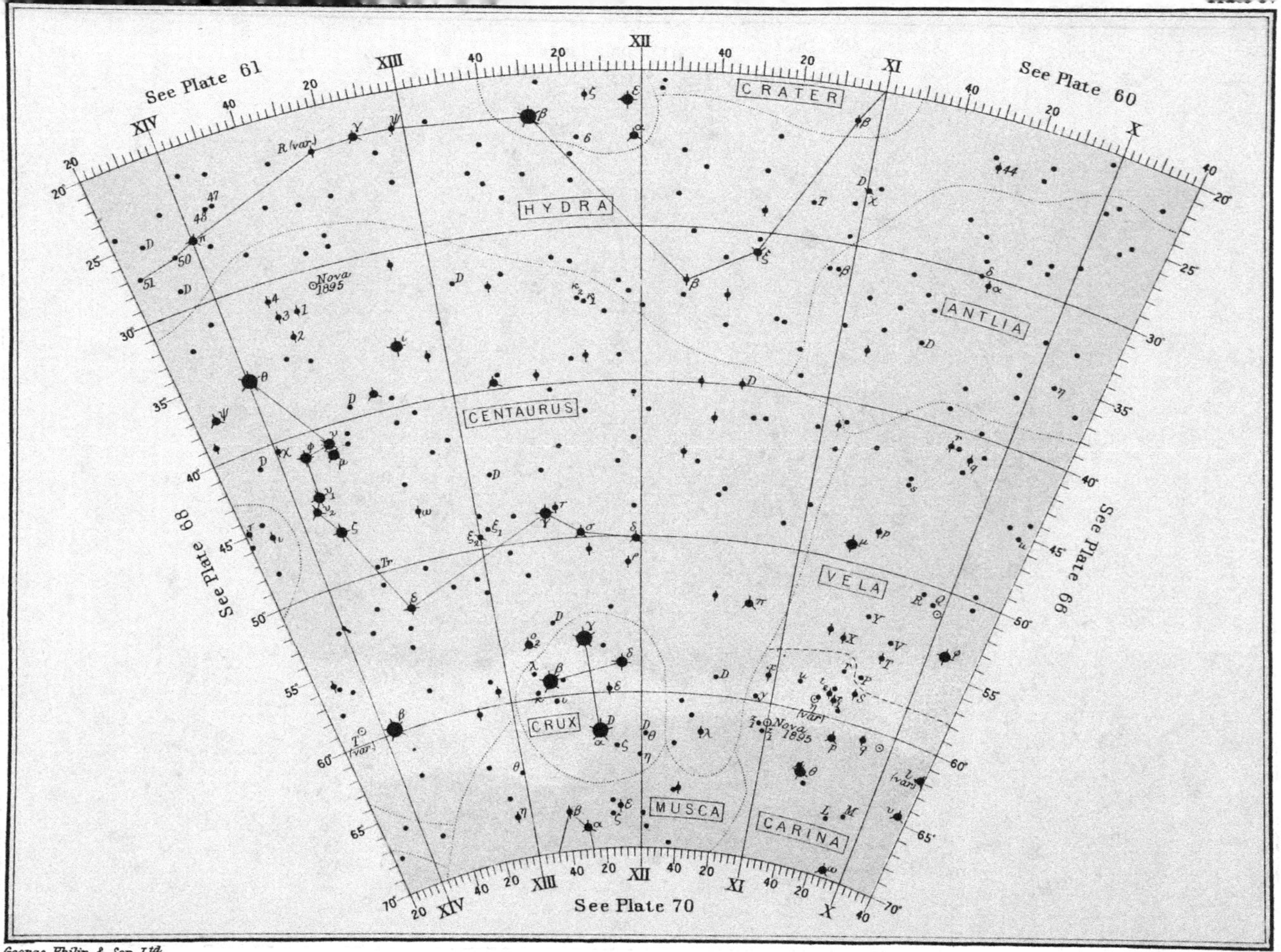

See Plate 60

See Plate 66

See Plate 61

See Plate 68

See Plate 70

CRATER

ANTLIA

VELA

CARINA

HYDRA

CENTAURUS

MUSCA

CRUX

Nova 1895

R (var)

Nova 1895

T (var)

The London Geographical Institute.

George Philip & Son, Ltd.

Plate 68

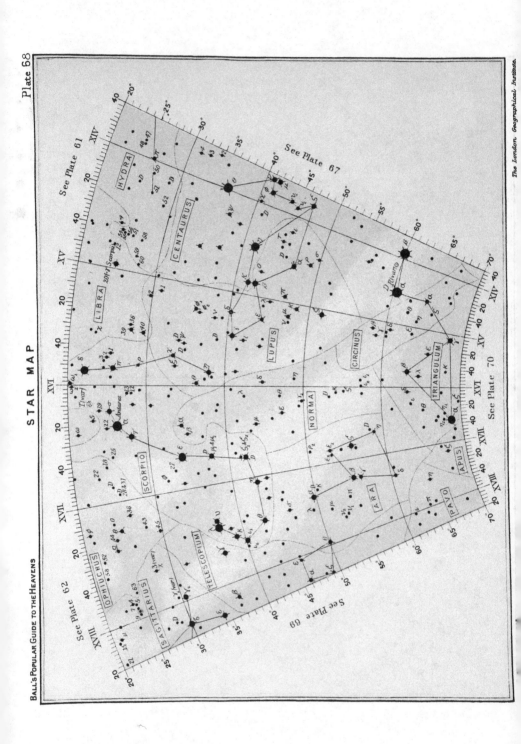

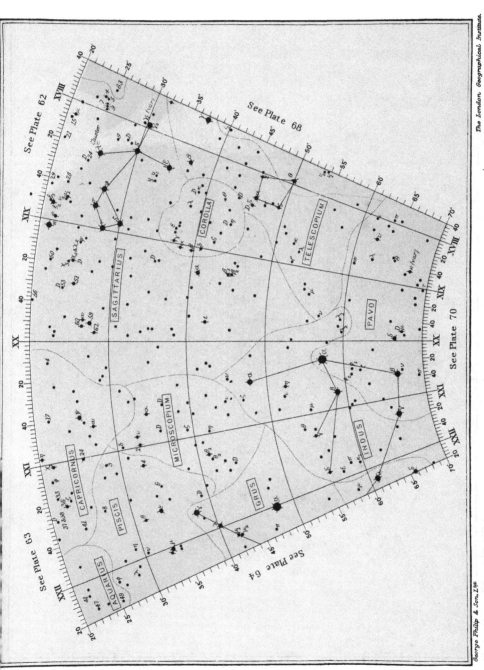

See Plate 62

See Plate 63

See Plate 68

See Plate 70

See Plate 64

SAGITTARIUS

COROLLA

TELESCOPIUM

PAVO

INDUS

GRUS

MICROSCOPIUM

PISCIS

CAPRICORNUS

AQUARIUS

Plate 70

STAR MAP

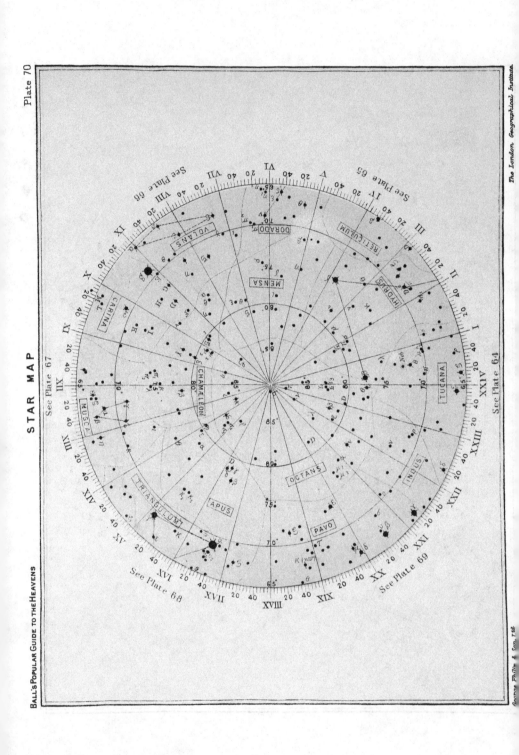

The London Geographical Institute.

George Philip & Son, Ltd.

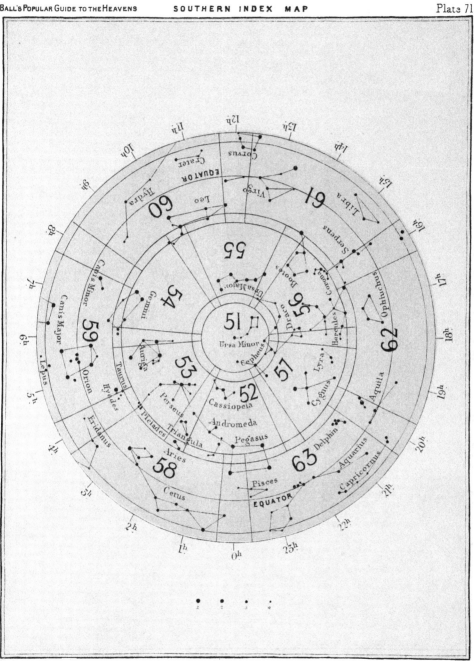

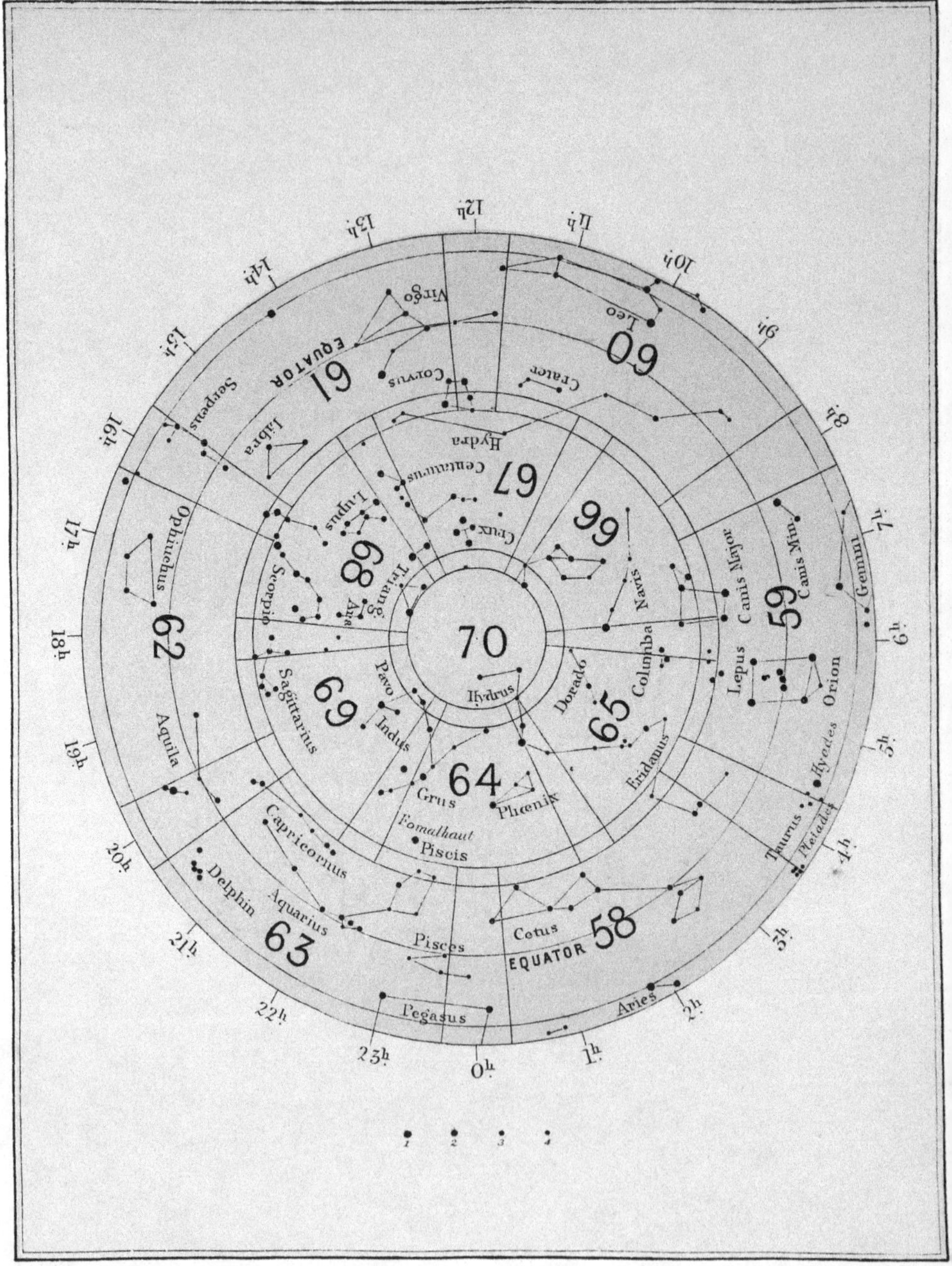

13ʰ 12ʰ 11ʰ 10ʰ 14ʰ

Virgo

Leo

60

13ʰ 9ʰ

Serpens EQUATOR 61 Corvus Crater

Libra Hydra 8ʰ

16ʰ Lupus Centaurus 67 66

Ophiuchus Scorpio 68 Crux 59 Canis Maj Canis Min 7ʰ

17ʰ Ara Triang Navis Gemini 6ʰ

62 70 Columba Lepus Orion 5ʰ

18ʰ Sagittarius 65 Dorado

19ʰ 69 Pavo Hydrus Eridanus Hyades Taurus 4ʰ

Aquila Indus Phœnix Pleiades

20ʰ 64 Grus

Capricornus Fomalhaut 58 Cetus 3ʰ

Delphin Piscis

21ʰ Aquarius Cetus EQUATOR 58

63 Pisces 2ʰ

22ʰ Pegasus Aries

23ʰ 0ʰ 1ʰ

·   •   •   •
1   2   3   4

George Philip & Son. Lᵈ   The London Geographical Institute.

CHAPTER VIII.

PLATE 73.

THE GREAT NEBULA IN ORION.

The Great Nebula which lies around the central star in the Sword of Orion is dimly visible to the naked eye as a blurring of the star disc. In the telescope it is a splendid object, full of intricate detail. But to bring out its full beauty we must, as always with the nebulæ, photograph it. It is shown then as an object of extraordinary complexity, but without any obvious plan. Enclosed in it are many stars which certainly belong to it, and are not merely seen by chance in the same direction ; for they share a typical spectrum whose characteristic is lines of helium which are strong also in the nebula itself.

The nebula shown in the plate is only the brighter and central portion of a much larger structure whose existence has recently been made clear, an immense spiral winding about the whole constellation of Orion. In a sense, then, the more familiar nebula is to be ranked with the spiral nebulæ, yet it is clearly distinguished from them by its spectrum, which shows that the light of the nebula comes from luminous gas, hydrogen, and helium, and the gas hitherto undetected upon Earth which has been called nebulium.

Of the real size and distance of the nebula practically nothing is known, beyond the fact that it is certainly immensely distant and large. And in this, as in many nebulæ, we find black holes with edges surprisingly sharp which are very hard to explain, except upon the highly speculative assumption that they represent dark material structures of some kind interposed between us and the shining nebula.

In this, as in all pictures taken with reflecting telescopes, the discs of the brighter stars are disfigured by rays which are of purely instrumental origin.

This plate, and the one that follows, were taken at the Yerkes' Observatory, Chicago, by Mr. Ritchey, with a reflecting telescope of two feet aperture made in the workshops of the Observatory. They owe a part of their beauty to the fact that the negatives have been reduced in the bright central portions which are inevitably over-exposed during the long time that is required to photograph the faint outlying portions of the nebula. Only by such a process is it possible to show on one plate the details both of the central and the remote parts of the nebula.

PLATE 74.

THE GREAT NEBULA IN ANDROMEDA.

In a telescope of moderate power this Great Nebula appears almost structureless, a soft elliptical mass of light steadily brightening towards the centre. In a large telescope, under exceptionally fine conditions, its outlying portions are seen to be furrowed with dark passages. But no eye has ever seen directly a hundredth part of the structure that is revealed by photography. Surrounding the central nucleus, which remains structureless, are a number of fainter rings, or perhaps more probably the convolutions of a spiral—it is almost impossible to say which—and the whole is mottled with brighter patches and perforated with dark holes of remarkable sharpness. The light of the nebula is such as would be given by vast numbers of stars crowded together—that is to say, the spectrum is continuous, the light is white. But in the present state of knowledge it is altogether premature to argue from that, that the nebula is really an enormous system of stars very far away. The most we can say is that its composition is certainly very different from that of the gaseous nebulæ.

PLATE 75.

THE GREAT STAR-CLUSTER IN HERCULES AND A NEBULA IN CYGNUS.

The photographs reproduced in this plate were taken by Mr. W. E. Wilson, at his observatory at Daramona, Westmeath, with a two-foot reflector.

Star-clusters may be roughly divided into two classes : loose and condensed. Of the former, the Pleiades cluster is the most conspicuous example. (*See* Plates 79 and 80.) The stars in it do not crowd towards the centre ; on the contrary, the density of stars in the centre of that cluster is less than the density in the surrounding sky ; but the individual stars are brighter. In a globular cluster, on the other hand—such as the Hercules cluster—small stars are so crowded towards the centre that, in all but the largest instruments, they become indistinguishable from one another.

PLATE 76.

THE SPIRAL NEBULA IN CANES VENATICI.

This, the most famous of the Spiral Nebula, had its true character first recognised by Lord Rosse with his great reflector at Parsonstown in Ireland. We are so happily situated with respect to it that we get a fair side view of it, and can trace in considerable detail how its branches are interlaced and studded with condensations which look as if they are on the way to become stars. Recent photographic work has shown that a large proportion of the nebulae, both known and hitherto unknown, are spirals, and this form must now be considered almost the rule instead of the exception.

This photograph, and the two following, were taken at the Lick Observatory by the late Professor Keeler, with the three foot Crossley Reflector, which was mounted at Ealing by the late Dr. Common, sold to Mr. Crossley, of Halifax, and afterwards presented by him to the Lick Observatory, in order that it might be worked under skies more favourable than those of England.

PLATE 77.

THE DUMB-BELL NEBULA.

It is a striking illustration of the power of photography in depicting nebulæ, that it has brought out a distinct resemblance between the Dumb-bell in Vulpecula and the Ring in Lyra, which could hardly have been suspected from the visual appearance of those objects. If we imagine the nebulosity, which exists inside the Ring, to shine a little more brightly, so that it fills up the Ring, and at the same time imagine the tendency towards thinning out at the ends of the longest diameter to be a little more pronounced, we shall see how easily the Ring might be transformed into the Dumb-bell. Both are gaseous, and both have a central star. It is difficult to resist the conclusion that the two nebulæ are closely related in kind.

PLATE 78.

THE RING NEBULA IN LYRA.

The Ring Nebula in Lyra can be easily found. It lies in the line from β to γ Lyræ, about one third of the distance from β to γ. It may be seen with a telescope of a few inches aperture, but it is doubtful if any telescope in the world, excepting perhaps Lord Rosse's reflector when in its finest condition, has ever shown to the eye so much as is presented in the photograph here reproduced, which was taken by Prof. Keeler at the Lick Observatory, with the Crossley Reflector.

The central star, which is so conspicuous on the photograph, is barely visible in the largest telescopes. It is much brighter photographically than visually, probably because its light is composed chiefly of those rays of short wave length to which the plate is sensitive but the eye nearly insensitive.

The photograph shows quite plainly that the ring is not uniformly bright; there are even some indications that it is composed of several interlacing or over-lapping rings, and it is remarkable how the ring thins out at the ends of its longest diameter. With longer exposures the centre of the ring fills up, and the nebula becomes a disk. It follows that the ring-like appearance is in a sense deceptive; that the real shape of the nebula is something like a hollow shell of gas, of which the borders look brighter, perhaps, because one is then looking through a greater depth of the shining matter : but this is at best a conjecture. What is certainly known is that the matter which shines is of the nature of self-luminous gas, giving a bright line spectrum. About the distance and the real size of the object very little is known, but it is practically certain that were our solar system placed in the centre of it, it would all lie within the space covered by the photographic image of the central star.

A fine example of a nebula with no central condensation is that in Cygnus. (No. .6992 in Dreyer's New General Catalogue.) Irregular and far-stretching nebulae such as this are not uncommon in the Milky Way. They are probably all gaseous, and seem to belong to a class altogether different from the spiral nebulae.

THE PLEIADES.

Plates 79 and 80 are representations of the Pleiades made with telescopes of different types. Plate 79 is from a photograph obtained by Dr. Isaac Roberts with his reflecting telescope of 20 inches aperture and 98 inches focal length—a ratio of aperture to focal length of about 1 : 5. Such an instrument is the most efficient kind of telescope for photographing faint nebulosity, and with similar instruments at different observatories the photographs of plates 73 to 78 have also been obtained. Plate 80 is engraved from a photograph made at the Paris Observatory by the Brothers Henry with the refracting telescope of 13 inches aperture and 135 inches focal length, which is the standard pattern in use at eighteen observatories cooperating in making the photographic chart and catalogue of the heavens. Such an instrument is not so suitable for photographing faint nebulae ; but it makes better photographs for measurement. And in presenting a series of the best *pictures* of some of the most beautiful objects in the sky, one must not fail to call attention to the other branch of astronomical photography, the making of plates which are not so pictorially effective, but which are more suitable for accurate measurement. Year by year the measurement of photographs replaces the older method of measurement at the telescope, and at the present time (1903) several observatories have nearly finished their share of the great International Photographic Catalogue of Stars mentioned above, which will include the places of about 2,000,000 stars.

The Pleiades cluster is the finest example of a loose cluster of bright stars intermixed with nebulosity. And in this cluster the nebulosity is of a very remarkable character. It takes the form in many places of long straight wisps connecting directly the brighter stars. In examining Plate 79 care must be taken not to confuse these with the eight symmetrical rays proceeding from each of the brighter stars, which are caused by an instrumental defect unavoidable in reflecting telescopes. Taking the key map to Plate 80 as a guide we may trace on Plate 79 the straight extension of nebula from Electra towards Alcyone, the ray proceeding from below Electra straight through the stars numbered 1 and 7, and the remarkable ray which runs from star No. 10 in a straight line through several stars above Alcyone and close below 24. Of a somewhat different character, yet still with a marked tendency to arrangement in straight lines, is the nebula which involves Maia and Merope.

Even apart from the evidence of the connecting nebula, there is little doubt that the Pleiades is a real cluster of bright stars, and not a chance gathering of stars seen nearly in the same direction but at very different distances from us. Of relative motion among the stars of the group there is little or none ; but the whole group is drifting together at the rate of about 9″ of arc per century past the other stars in the neighbourhood.

PLATE 81.

THE MILKY WAY, AROUND THE STAR-CLUSTER MESSIER II.

This is a reproduction of one of the celebrated photographs of the " star clouds " of the Milky Way, taken by Professor Barnard at the Lick Observatory in 1889. His description of it is as follows : "The small cluster, Messier 11, lies on the upper or north edge of the neck of the large cloud, and looks like a nucleus. The western side of the great cloud has several rather sharply-marked indentations and several detached masses of stars. The star β *Aquilae*, on the upper north edge of the great head, has two curious sprays of stars extending from it, giving the appearance of a ram's horns. The great star-cloud seems to be made up of very small stars, apparently very uniform in size. Near the lower right-hand corner of the plate is shown a beautiful bright nebulous star. . . . . The nebulosity about this star is somewhat elliptical. It was discovered on the plates of 1899, and is quite noticeable visually. The bright star near the N.E. edge of the plate is λ *Aquilae*. The great star-cloud seems to stretch out to and surround this star."

No less striking than the brilliant clouds of stars are the dark holes and lanes which pierce them. These sharply-defined vacuities are characteristic features of the star-clouds, and they give some cause for suspicion that there may exist in space regions of light-stopping material which cut out the light of the stars beyond. It is difficult otherwise to account for the existence of so many well-defined empty spaces in a field of stars otherwise so rich.

PLATE 82.

NOVA PERSEI AND THE NEBULA IN MOTION.

The appearance of a new star in Perseus was first observed by Dr. Anderson at Edinburgh, on 1901, Feb. 21, at G.M.T. 14h. 40m. The star was then of magnitude 2 7. On the previous night a photograph had been taken by Mr. Stanley Williams at Brighton which showed stars down to the 12th magnitude, but no trace of the Nova. It is, therefore, certain that in little more than 24 hours the star must have increased in brightness more than ten thousand times. On Plate 82 we have the photograph above mentioned ; and, for comparison with it, a second photograph taken after the appearance of the star. So promptly was the discovery made and the news circulated that on the evening of February 22nd the star was under observation all over the northern hemisphere, and it was found that the brightness was still increasing. But in a few days it began to fail rapidly—though with very strongly-marked fluctuations—until by midsummer it was invisible to the naked eye while its spectrum, as is usual in such cases, had become that of a gaseous nebula.

In spite of the enormous amount of information which was given by this outburst—unsurpassed since the days of Tycho Brahe—it does not seem that we are much nearer an understanding of its cause. But one thing seems to be clear—the outburst of a new star is not due to the collision of two dark bodies which are thereby raised to a transcendant heat.

Almost more remarkable than the star itself was the nebula which was discovered around it in the autumn of 1901. The first satisfactory photograph of it was obtained at the Yerkes Observatory on September 20th. On November 7th and 8th a photograph obtained at the Lick Observatory showed that parts of the nebula were in rapid motion ; and the same thing was found independently at the Yerkes Observatory on the 9th and 13th. In the lower portion of our plate diagrams drawn to scale from the original negatives show these changes unmistakeably. If the pointed structures, lettered *a* and *e*, are compared on the two diagrams, and reference made to the surrounding stars and the scale at the side, it will be seen at once that these points have moved about a minute of arc in six weeks. Such a rate of motion is unprecedented, and many theories have been advanced to account for it. The most generally accepted theory is that there was around the star a complicated nebula too faint to be photographed until it was lit up by the burst of light which proceeded from the star; and, that the motion which was observed was not a real motion of the nebula itself, but the effect of successive lighting-up of different parts of the nebula as the light passed outwards over it. It is scarcely possible otherwise to account for a motion which must have been at least very nearly equal to the velocity of light itself.

0   10   20   30   40   50   60

S

E

N

HE GREAT NEBULA IN ANDROMEDA. (M. 31)   G. W. RITCHEY,   2-ft. Reflector, YERKES OBSERVATORY.

Scale : Minutes of Arc. Plate 74.

THE GREAT NEBULA IN ANDROMEDA. (M. 31) G. W. RITCHEY, 2-ft. Reflector, YERKES OBSERVATORY

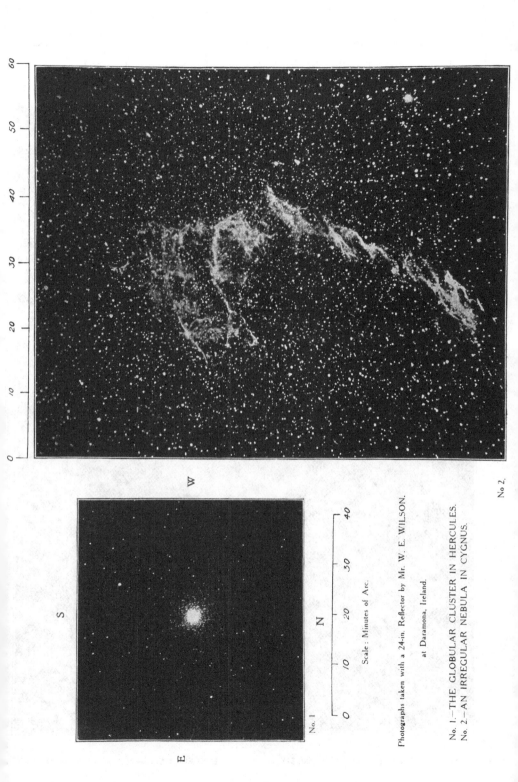

Scale: Minutes of Arc.

Photographs taken with a 24-in. Reflector by Mr. W. E. WILSON.

at Daramona, Ireland.

No. 1.—THE GLOBULAR CLUSTER IN HERCULES.
No. 2.—AN IRREGULAR NEBULA IN CYGNUS.

0   1   2   3   4   5   6   7   8   9   10

S

W

N

SPIRAL NEBULA IN CANES VENATICI. (M. 51)   J. E. KEELER, Crossley Reflector, LICK OBSERVATOR

Scale : Minutes of Arc.

S

E

N

THE RING NEBULA IN LYRA. (M. 57.)   J. E. KEELER, Crossley Reflector, LICK OBSERVATORY.

0   5   10

THE DUMB-BELL NEBULA IN VULPECULA   J. E. KEELER, Crossley Reflector, LICK OBSERVATORY

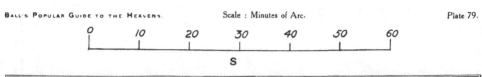

THE NEBULÆ IN THE PLEIADES.   ISAAC ROBERTS, 20-in. Reflector, Crowborough, Sussex.

THE PLEIADES

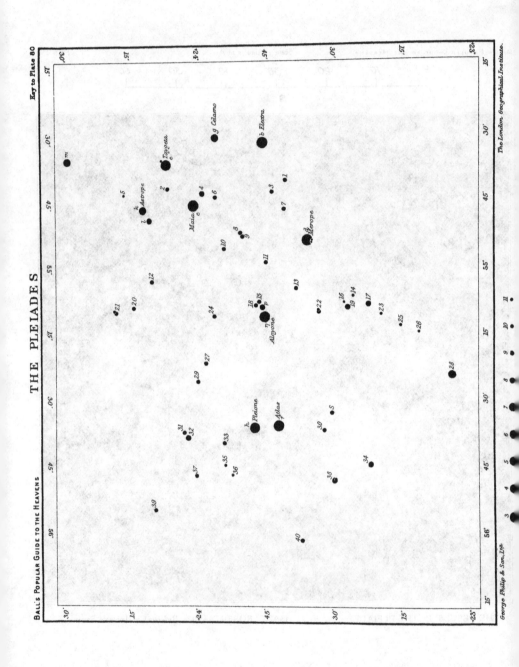

BALL'S POPULAR GUIDE TO THE HEAVENS

The London Geographical Institute.

George Philip & Son, Ltd.

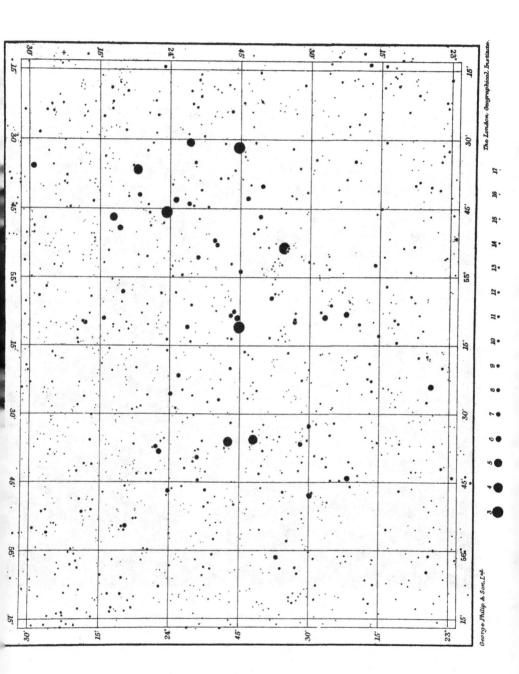

The London Geographical Institute.

George Philip & Son, Ltd.

Plate 81.

0   1   2   3   4   5   6   7   8   9   10

Scale : Degrees.

N

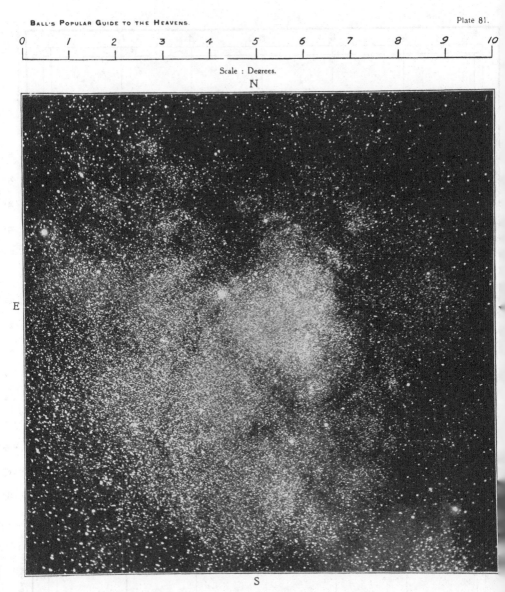

E

S

THE MILKY WAY AROUND THE STAR CLUSTER MESSIER II E. E. BARNARD, 6-in. Portrait Lens, LICK OBSERVAT

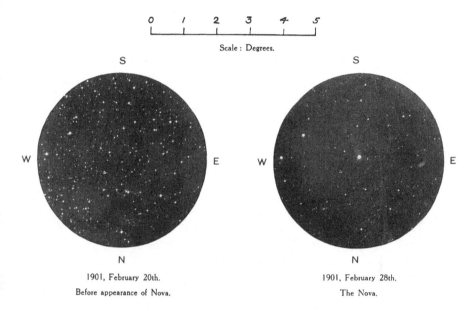

Scale : Degrees.

1901, February 20th.   1901, February 28th.
Before appearance of Nova.   The Nova.

NOVA PERSEI. Photographs by A. STANLEY WILLIAMS, Hove, Sussex.

Scale : Squares are Two Minutes of Arc.

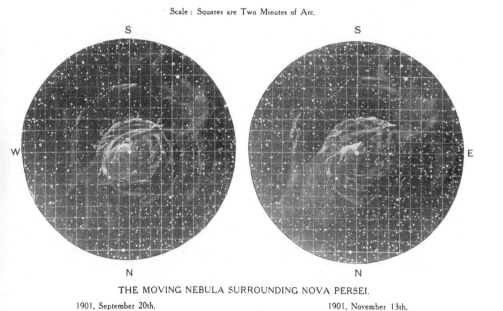

THE MOVING NEBULA SURROUNDING NOVA PERSEI.

1901, September 20th.   1901, November 13th.

Drawn by G. W. RITCHEY, from Photographs taken with the 24-in. Reflector, YERKES OBSERVATORY.

A SELECT LIST

OF

STARS, STAR CLUSTERS, AND NEBULÆ.

CHAPTER IX.

A SELECT LIST OF STARS, STAR CLUSTERS, AND NEBULÆ.

In preparing a list of objects which are suitable for observation with small instruments the author naturally turns in the first place to Admiral Smyth's "Celestial Cycle" for suggestions; and for the results of the most modern work upon those objects to Miss Clerke's "Problems in Astrophysics," Dr. See's "Evolution of the Stellar Systems," and Prof. Simon Newcomb's "The Stars : A Study of the Universe"; to the "Companion to the Observatory," the publications of the Royal Astronomical Society, and the scientific journals. The student will find more extensive lists of interesting objects in Mr. Gore's "The Stellar Heavens."

An attempt has been made in these notes to give some of the most recent results for the distance, mass, &c. of the star systems. A very brief summary of the principles from which these results have been deduced is here given :—

Distance of the Stars.

Astronomers find it convenient to express the distance of a star from the solar system by a quantity which is the apparent angular radius of the Earth's orbit as seen from the star; this quantity is called the annual parallax of the star. For example, the annual parallax of α Centauri, our nearest stellar neighbour, is 0″·75 ; that is to say, the radius of the Earth's orbit, as viewed from the star, would subtend an angle of three-quarters of a second of arc. Since a unit of length viewed at a distance of 200,000 units makes an angle of a second of arc very nearly, we have the following convenient rule :—

To find how many million times farther away than the Sun any given star is :—

Take the number expressing its annual parallax in seconds of arc ; multiply it by five ; and find the reciprocal of the product. This is the number of millions required

For example : The parallax of Capella is 0″·09. The reciprocal of 5 times 0·09 is about 2¼. Hence Capella is about 2¼ million times as far away as the Sun.

Distance of Star in Light-Years.

We have seen that astronomers generally express the distance of a star in terms of its annual parallax, always a very small angle, which decreases as the distance of the star increases. It is inconvenient to try to express the distances in the ordinary astronomical unit of length, the distance of the Earth from the Sun ; the numbers are too large. But the distance which light travels in one year makes a unit of convenient size. Thus, since according to the best determination, the time taken by light to travel the distance R from the Earth to the Sun is 498·46 secs., in one year (365d. 5h. 48m. 46s.) light travels about 63,300 R.

This is called a light-year.

And a star which has a parallax of 1″ is distant 206,265 R.

Hence light would take 3·26 years to come from that star to the Earth.

And for any other star whose parallax expressed in seconds of arc is given, the time taken by light to come from it to our system is $\dfrac{3·26}{\text{parallax}}$.

Example : The parallax of Capella is 0″·09. Hence light takes $\dfrac{3·26}{0·09}$ = about 36 years on its journey.

Velocity of Star at Right Angles to the Line of Sight.

This can be found when the star's parallax and proper motion are known. If the star has a parallax of 1″ and a proper motion of 1″, it moves during the year a distance at right angles to the line of sight equal to the radius of the Earth's orbit, that is, about 92,900,000 miles. This corresponds to a speed of 2·94 miles per second. If for any star we multiply this number by the annual proper motion, and divide by the parallax, we get the velocity of the star at right angles to the line of sight.

Example : α Lyræ. Proper motion 0″·36 ; parallax 0″·08. Velocity at right angles to line of sight = $\dfrac{0·36}{0·08}$ × 2·94 = about 13 miles per second.

Brightness of Stars Compared with the Sun.

To give an account of the methods of comparing the relative brightness of the Sun and stars would be beyond the limits of the present work. But the formula which represents the relation is comparatively simple. If π is the parallax of the star, m its magnitude, and r the ratio of its light to the light of the Sun removed to the distance of the star

$$\log r = \frac{4}{10}\left(5 \log \frac{1}{\pi} - m.\right).$$

Example : For Procyon we have mag. 0·47, parallax 0″·33 ; whence log. r = 0·77, r = 5·9

It should be noticed that estimates of this kind are very much affected by small changes in the adopted value of the parallax of the star, and are therefore necessarily rather uncertain.

Masses of the Stars Compared with the Sun.

Can be found only for binary stars.

When a binary star has completed enough of a revolution to make it possible to compute all the circumstances of its apparent orbit, it is possible to compute the shape and position of the real orbit, which will differ from the apparent orbit, unless it happens to lie square to the line of sight ; we also know its size in seconds of arc. If, in addition, we know the parallax, i.e., the distance of the star, we know the length of the major axis of the real orbit in terms of the distance of the Earth from the Sun.

For example : The major axis of the real orbit of the companion of Sirius about that star subtends to us an angle 8″·03 ; the major axis of the Earth's orbit round the Sun subtends at the distance of Sirius the angle $0''\cdot37$ (the annual parallax). Hence the distance of its companion from Sirius is $\dfrac{8\cdot03}{0\cdot37}$ times, $= 21\cdot7$ times the distance of Earth from Sun. The companion of Sirius revolves in $52\cdot2$ years. It follows from an extension of Kepler's 3rd law, that mass of Sirius + companion $= \dfrac{(21\cdot7)^3}{(52\cdot2)^2}$ mass of Sun + Earth. The mass of the Earth is negligible compared with that of the Sun, and we have, therefore, on reducing the fraction, Mass of Sirius + companion $= 3\cdot7$ times mass of Sun.

Spectroscopy.

Without entering into the complicated theory and methods of spectrum analysis, it is possible to indicate broadly the facts upon which spectroscopic determinations of various kinds are based.

When the source of light is transparent glowing gas, the spectrum consists of a number of isolated bright lines.

When the source of light is a glowing solid body, the spectrum consists of a continuous rainbow band of colour, with no details at all. It is visible from the red to the violet; below the red it can be detected by its heating effects; beyond the visible violet it continues for some distance to affect the photographic plate.

If the hot solid body is overlaid by cooler layers of gas, dark absorption lines appear in the continuous spectrum, in the exact places where bright lines would be seen were that gas shining alone.

Upon this fact rests the possibility of determining what substances are present in the vapours surrounding the Sun and stars, and in the nebulæ. If iron, for example, is vapourised in the electric arc its spectrum consists of bright lines.

In the spectrum of the Sun are found dark lines corresponding exactly in position with each one of these bright lines. The conclusion is that iron exists in a state of vapour above the Sun, relatively cooler than the glowing solid particles in the photosphere below. In a similar way the presence of a great number of other elements is detected in the Sun and stars.

Again, if hydrogen is made to glow electrically in a vacuum tube its spectrum consists of certain bright lines. In the spectrum of a gaseous nebula bright lines are found in the same positions. It follows that hydrogen is shining in the nebula.

But the comparative positions of these lines remain fixed only so long as the sources of light are at rest with respect to one another. If a star is in motion with respect to the Earth all the lines of its spectrum are slightly shifted—towards the violet if the star is approaching, towards the red if it is receding. By measuring the amount of this shift it is possible to decide what is the velocity of approach or recession of the star.

Spectroscopic Binaries.

The velocities of a great number of stars relative to the solar system have been measured by this method. In a great many cases the velocity has been found to vary regularly in a definite period. The conclusion is that the star is in orbital motion round the centre of gravity of itself and a companion.

In the majority of cases the companion does not give enough light to affect very much the spectrum of the principal star. In some cases, however, the two stars are nearly of the same brightness. Then one star will necessarily be approaching while the other is receding, and *vice versâ ;* and the spectrum will be doubled, each star will show its own set of dark lines, which will alternately close up on and open out from each other, and in such cases the duplicity of the star is obvious, without need of reference to the positions of comparison lines obtained terrestrially.

If the velocities in their orbits of the components of a binary are known, and the period of revolution, it is possible to calculate the real size of the orbits, and thence the masses as above. In the case of spectroscopic binaries one cannot usually solve the problem completely, but can determine that the mass of the pair must be at least as much as a certain quantity.

---

35 Piscium. 0h. 10m. + 8° 13′.

A fine, double star, 6th magnitude white, and 8th magnitude purplish. The components appear to be relatively fixed, in position angle 150°, and distance 12″.

Globular Cluster 47 Toucani. 0h. 20m. - 72° 39′.

A magnificent cluster containing about 1,500 stars within a radius of about 3′. Visible to the naked eye as a hazy star, of light equivalent to 4½ magnitude. The cluster contains six variable stars.

Nebula M. 31 Andromedæ. 0h. 37m. + 40° 43′.

One of the most splendid nebulæ in the sky, but not very interesting as a telescopic object. The best way to see nebulæ in the telescope is to set the instrument just ahead of the nebula and allow it to drift into the field. If close attention is paid it is possible to see in the Andromeda nebula certain dark and apparently straight lanes ; otherwise the nebula appears almost structureless, fading away gradually from the bright centre. Its real complicated structure can only be seen well in the photographs (*see* Plate 74), where it appears as a fine spiral seen obliquely with a great deal of curious detail. Its spectrum is continuous, and dark lines are suspected in it ; it is not, therefore, it would seem, one of the transparent gaseous nebulæ.

η Cassiopeiæ. 0h. 43m. + 57° 17′.

A binary star of rapid motion, and large parallax. Magnitude 3½ and 7½ ; distance 5″·68 ; position-angle 226°·4 (Maw, 1903·2). Its period is 196 years, and its parallax is 0″ 20. From these data it may be concluded that the mass of this pair of suns is 1·8 times the mass of our Sun, that their luminosity is together about equal to that of the Sun, and that their mean distance apart is 41 times the distance of the Earth from the Sun. But the caution should be given that all such deductions may be considerably modified by a small change in the value of the parallax adopted ; and the numbers must be taken as examples of the kind of information that these researches will give us, rather than as absolutely determined quantities.

a Ursæ Minoris (Polaris). 1h. 23m. + 88° 46′.

The Pole Star is the best known and most practically important star in the sky. On account of its proximity to the North Pole of the sky it appears to the eye to be almost devoid of the ordinary daily movement of the stars about the Pole. The actual diameter of the small circle which it describes daily is 2° 25′ (1903), about five times the apparent diameter of the Moon. The Pole Star can easily be found by the aid of the "pointers," α and β Ursæ Majoris. It is a standard 2nd magnitude star, with a small 9th magnitude companion distant 19″ in position-angle 212°.

It has been shown quite recently by Campbell that the Pole Star is a spectroscopic binary, with a period of very nearly four days, and a slow orbital motion of four miles per second. But irregularities have been found in this motion, and it seems probable that Polaris has *two* dark companions.

γ Arietis.   1h. 48m. + 18° 49′.

This star is interesting as having been discovered as a double star by Hooke, as early as 1664, when he was observing the Comet of that year—"a like instance to which I have not else met with in all the heavens." Magnitude 4·2, 4·4 ; distance 8″·3 ; position-angle 358°. Easily visible with a small telescope.

α Piscium.   1h. 57m. + 2° 17′.

A fine double star ; components about 3 and 4 magnitude ; distance 3½″ ; position-angle 359°.

γ Andromedæ.   1h. 58m. + 41° 51′.

One of the finest double stars in the sky. Magnitude 2½, yellow, and 5½, blue green. Distance 10″·2 ; position-angle 62°. The blue star is itself a binary, distance 0″·45 (1903), a difficult object at present in any telescope less than 12-in. aperture. The period of this small binary is 55 years, and the eccentricity of its orbit is very great. In 1890 the stars were only two or three-hundredths of a second apart, and no telescope could separate them.

ι Trianguli.   2h. 7m. + 29° 50′.

An exquisite double star, of which the primary is yellow, 5th magnitude, and the companion blue, 7th magnitude. Distance about 3″·5, and position-angle 75°. During the 80 years in which the star has been under observation there has been a slight diminution in position-angle and increase in distance, so the star is probably a very slow binary.

The Star Clusters in Perseus.   2h. 12m. + 56° 41′.

A splendid pair of clusters of bright stars, visible to the naked eye as a bright patch in the Milky way, on the line joining α Persei with δ Cassiopeiæ, at about three-fifths of the distance from the former. The preceding cluster contains two bright stars of the 7th magnitude and a beautiful "horse-shoe" of 9th and 10th magnitude stars. The cluster which follows about 3m. on the same parallel is not so fine, but contains two conspicuous triangles of stars.

o (Mira) Ceti.   2h. 14m. − 3° 26′.

A very remarkable variable star discovered three hundred years ago by the German astronomer Fabricius, who was the first observer of sun spots with the telescope. It varies in a period of 330 days, more or less, from about the 3rd magnitude (on the average) down to 9½ ; it is impossible to define its behaviour accurately since no two successive cycles are similar. The student will readily find its place with reference to other stars from the charts, and will find great interest in observing its variations.

θ **Persei.**   2h. 37m. + 48° 49'.

An interesting triple star—A of the 4th, B and C of the 10th magnitudes. A and B are affected with the same proper motion, amounting to nearly 1" annually, and probably form a binary system. Distance 17"·4, position-angle 299° (1900). C is at distance 80", and position-angle 225° (1900), and does not share the proper motion of the other two, so that they are rapidly separating from it.

γ **Ceti.**   2h. 38m. + 2° 49'.

A beautiful double star, A 3·0 m. yellow, and B 6·8 m. b ue, with common proper motion. Probably a slow binary. Distance 3"·5 ; position angle 292°. (1899).

θ **Eridani.**   2h. 55m. - 40° 41'.

A fine double star for southern observers, magnitudes 3·5 and 5·5 ; distance 8"·2 ; position angle 85° It is practically certain that this pair, now a star of the 3rd magnitude to the naked eye, is identical with Ptolemy's "Last of the river" and with the Achernar of Al-Sufi, who describes it as of the first magnitude. This is one of the clearest cases of a star having lost a large percentage of its light within historical times.

β **Persei (Algol).**   3h. 2m. + 40° 34'.

The most famous variable star in the sky. Every 2d. 21h. its light suddenly begins to diminish from magnitude 2·4, until in a little over 4 hours it has sunk to 3·6 ; without any appreciable pause it then rises again in nearly the same time to its normal brightness. The conjecture that this is due to an eclipse by a dull companion has been confirmed by the spectroscope. Algol is thus a spectroscopic binary whose plane passes nearly through the solar system. This being determined we have the following data : Velocity of Algol in orbit 26 miles per second ; radius of orbit 1,000,000 miles. Diameter of bright star about 1,000,000 miles ; of dull companion about 800,000. In default of a knowledge of the mass of the companion we cannot determine the separation in miles of the two stars, but they must be relatively very close together, only a few million miles apart. If we assume that the two stars have the same density, their masses are nearly in the ratio of 2 : 1, and their distance apart about 3¼ million miles. They, like all eclipsing variables, seem to be a great deal less dense than the Sun.

The Pleiades. Central Star, Alcyone.   3h. 41m. + 23° 48'.

The well known naked eye cluster of bright stars, seen to greatest advantage in a telescope of low magnifying power. Most of the brighter stars have a common proper motion, and form, without doubt, a real group. Although the cluster is so rich in bright stars it contains actually fewer fainter stars than equal areas of the surrounding sky. The cluster is full of nebulosity which has an apparent connection with the stars. (*See* Plate 79).

α Tauri (Aldebaran).   4h. 30m. + 16° 18′.

This star, Mag. 1·2, conspicuous for its ruddy colour, is the principal object in the group of the Hyades. (Map 59). Its proper motion is 0″·19 and parallax 0″·12, whence its light is about 23 times that of the Sun, and its motion across the line of sight about 4 miles per second. An occultation of Aldebaran by the Moon, which not infrequently occurs, is a striking phenomenon.

α Aurigæ (Capella).   5h. 9m. + 45° 54′.

One of the brightest stars in the sky. Newall and Campbell found independently that it is a spectroscopic binary, components unequally bright, but nearly equal in mass, moving in orbits of radius about 50,000,000 miles in a period of 104 days ; joint mass about seventeen times Sun. The parallax of the stars has been carefully determined as 0″·08 ; and this implies that the stars give together about 130 times as much light as the sun ; they must therefore be much brighter mass for mass. From the data just given it is easy to see that they might be seen telescopically as a double star with distance 0″·1, and the Greenwich observers believe that they have seen the star elongated though not clearly divided.

Proper motion 0″·43 ; annual parallax 0″·08, whence its velocity at right angles to the line of sight is about 16 miles per second.

Nebula M. 1 Tauri.   5h. 29m. + 21° 57′.

The " Crab " nebula, so called by Lord Rosse because of the claw-like protuberances which he observed. In a small telescope it is not very interesting ; but it is famous as the object which induced Messier to draw up his celebrated catalogue of nebulæ, by the numbers of which the brighter nebulæ are known to this day. The Crab Nebula is No. 1.

θ Orionis.   5h. 29m. − 5° 28′.

A splendid multiple star involved in the brightest part of the great nebula in Orion. Four bright stars, of magnitude 6, 7, 7½, and 8, are the well-known " trapezium." There are a number of fainter stars included in the group.

σ Orionis.   5h. 34m. − 2° 39′.

A very fine multiple star. In small telescopes it presents the appearance described by Sir William Herschel of " a double-treble star, or two sets of treble stars almost similarly situated." Larger instruments show a number of other stars, and Burnham has found that the brightest star is itself double, and a rapid binary.

ζ Orionis.   5h. 36m. − 2° 0′.

This, the following of the three stars in Orion's belt, is a fine double, with a distant faint companion. The components of the double are of magnitude 2 and 6 ; distance 2″·4 ; position-angle 156°, slowly increasing.. The faint companion, magnitude 10, is in position-angle 9°, at distance 57″.

Cluster M. 37 Aurigæ. 5h. 46m. + 32° 31′.

A magnificent cluster of small stars, loose and little condensed. It does not appear that there is any nebulosity in this cluster, though in small instruments the crowd of small stars presents the appearance of it. " Even in smaller instruments extremely beautiful, one of the finest of its class. Gaze at it well and long."— *Webb.*

a Orionis (Betelgueuse). 5h. 50m. + 7 23′.

A bright yellowish-red star, whose light is somewhat variable, about 0·9 usually. Like nearly all stars in the constellation of Orion it has a small proper motion, 0″·027 per annum, and a small parallax, 0″·024. It follows that this star gives several hundred times the light of the Sun, but its motion across the line of sight is slow, about 3 miles per second.

β Aurigæ. 5h. 52m. + 44° 56′.

Telescopically a single star. But the lines in its spectrum appear alternately double and single every 48 hours, and the displacement indicates a relative velocity of 150 miles a second. It follows that the star is double, with components nearly equally bright, revolving in a period of just less than 4 days. The orbit is somewhat eccentric ; the stars are at least 7,500,000 miles apart, and their combined mass 4½ times that of the Sun.

Star Cluster M. 35 Geminorum. 6h. 3m. + 24° 21′.

A fine and bright, but loose cluster of stars, without much trace of the condensation towards the centre which characterizes a globular cluster.

11 Monocerotis. 6h. 24m. − 6° 57′.

A very striking triple star, A of the 5th magnitude, B and C of the 6th. B is in position angle 131° and distance 7″ ; C in 120°, distance 9½″. There is no evidence of relative motion in this system.

a Canis Majoris (Sirius). 6h. 41m. − 16° 34′.

The brightest star in the sky, Mag. − 1·6, of large proper motion, and large parallax. Irregularities in its proper motion suggested that the star must have a companion, which was discovered in 1862. Its magnitude is about 9, but it is visible only in the largest telescopes being very hard to see on account of its nearness to the brilliant primary. In 1903·1 its distance was 6″·3, in position angle 128°. The parallax of Sirius as determined by Gill and Elkin is 0″·37. The period of the companion is 52 years, and mean distance from Sirius 8″·03· From these data we may conclude that the total mass of the pair is 3·7 times that of the Sun, but their combined light is 32 times ; that their distance apart is 22 times that of the Earth from the Sun. From the irregularity in the proper motion of Sirius it may further be, shown that Sirius is only about twice as massive as its companion, though it is 10,000 times as bright. The great intensity of the light of Sirius, 30 times that of the Sun, with only 2½ times its mass, and the dimness of the companion are very remarkable.

a Geminorum (Castor). 7h. 28m. ÷ 32° 6′.

A very fine double star, one of the best objects for small telescopes. Magnitude 2·0 and 2·8 ; distance (1902) 5″·7 ; position angle 223°. Period of revolution about 1,000 years. The interest of this system has been greatly increased by the discovery that the fainter component is a spectroscopic binary with a heavy dark companion. Period 2 95 days ; velocity of bright star in orbit 22 miles per second ; radius of orbit at least 1,800,000 miles.

a Canis Minoris (Procyon). 7h. 34m. + 5° 29'.

One of our nearer neighbours among the stars. Annual parallax 0"·33 ; proper motion 1"·25 ; Mag. 0·47, whence its light is about six times that of the Sun, and its velocity at right angles to the line of sight about 11 miles a second.

A binary star with a faint but relatively very massive companion, whose presence first became known by the large irregularities which its attraction produces in the motion of the principal star. The disturbing companion was at last discovered with the great Lick telescope in 1895. Its mass is about equal to that of the Sun, but the light that it gives is very much less, perhaps about one-thousandth.

ζ Cancri. 8h. 6m. + 17° 59'.

One of the most remarkable *multiple* stars in the heavens. It is composed in the first place of two stars, A and B, of the 5 and 5·7 magnitude respectively, whose orbit has been well determined. These two revolve around each other, in a period of 60 years, at a distance of less than 1", and are accompanied by a third star, C, of 5·5 magnitude, which revolves around the centre of gravity of all in an opposite direction. From irregularities in the motion of C, which take place in a period of 17½ years, it i concluded that it is but a satellite of an invisible body around which it revolves in that time, describing an ellipse with a radius of about one-fifth of a second, and that the two together circle around A and B in 600 or 700 years.

Cluster M. 44 Cancri. 8h. 34m. + 20° 21'.

A large and loose cluster of stars known as Præsepe, or the Bee-hive. To the naked eye it appears as a nebulous patch of light a little south preceding γ Cancri. A fine object in small telescopes.

ε Hydræ. 8h. 42m. + 6° 50'.

A beautiful triple star. A and B are respectively of the 4th and 6th magnitude, and are so close that only the most powerful telescopes can separate them. Position-angle 23°, distance 0"·13 (1902), yellow. The companion C is 7th magnitude, blue, in position-angle 234°, distance 3"·47 (1902).

a Leonis (Regulus). 10h. 3m. + 12° 27'.

This bright star (magnitude 1·23) has quite a large proper motion, 0"·27 per annum, but a small parallax, 0" 02, whence it follows that its light must be equal to that of 1,000 of our Suns.

γ Leonis. 10h. 14m. + 20° 22'.

A very fine double star, orange yellow. Magnitude 2 and 4 ; distance 3"·81 ; position angle 115°·7 (1903.8, Lewis). Binary with a period of about 400 years.

Planetary Nebula H. IV. 27 Hydræ.   10h. 20m. — 18° 8'.

A typical planetary nebula, whose light is equal to that of an 8th magnitude star. Admiral Smyth describes it as "resembling Jupiter in size, equable light, and colour," though of course it is not nearly so bright. Its spectrum consists of bright lines, and it is therefore gaseous.

η Argus.   10h. 41m. — 59° 10'.

This, one of the most remarkable stars in the sky, set in the middle of one of the most remarkable nebulæ, is unfortunately too far south to be visible in European latitudes. During the 18th and early part of the 19th centuries it was a naked-eye star between the 2nd and 4th magnitude. In 1837 it rose quickly in brightness to first magnitude, faded a little, and in 1843 rose very nearly to the brightness of Sirius. In the following 30 years it sank steadily to magnitude 7½, where it remains. Its spectrum is of the peculiar type associated with the temporary stars, and it seems to differ from them principally in being semi-permanent.

Planetary Nebula M. 97 Ursæ Majoris.   11h. 9m. + 55° 34'.

In small telescopes a faintly luminous disc about the size of Jupiter. In very large telescopes it appears to have a very complicated structure. The Earl of Rosse found two condensations surrounded by spirals in opposite directions, from which it obtained the name of the "Owl Nebula." This, like nearly all planetary nebulæ, gives a spectrum of bright lines, and is therefore gaseous.

ξ Ursæ Majoris.   11h. 13m. + 32° 6'.

A beautiful double star, rather close for small telescopes. Magnitude 4 and 5 ; distance 2".3 ; position-angle 144° (1902). Period about 60 years. The brighter component has been shown to have a variable velocity in the line of sight, which shows that there is a third star in the system, but the data are not yet complete.

This star was one of the first stars recognized as binary, having components which move about their common centre of gravity in accordance with the law of universal gravitation, and it was actually the first whose orbit was computed on gravitational principles.

ι Leonis.   11h. 19m. + 11° 5'.

A rather close double star. A, 4th magnitude, pale yellow, and B, 7½ magnitude, blue. Position-angle 53° ; distance 2".17 (1900). There is considerable relative motion, and it is almost certainly a binary.

24 Comæ.   12h. 30m. + 18° 56'.

A fine, but wide, double star. A, 5½ magnitude, orange, and B, 7 magnitude, blue. Position-angle 271° ; distance 20". The colours form " a striking and beautiful contrast."

α Crucis.   12h. 21m. — 62° 33'.

This, the brightest star in the Southern Cross, is a very fine triple star. Magnitudes 1·5, 1·8, and 6 ; the bright stars a fairly close pair, distance 5".0, position-angle 118° ; and the fainter, distant 90" in position-angle 202°. The parallax of the bright stars is 0".05.

γ Virginis. 12h. 37m. − 0° 54′.

A most interesting binary system, consisting of two stars of the 3rd magnitude ; distance 5″·74 ; position-angle 328°·1 (1902). Period about 194 years. Its orbit is very eccentric. In 1831 the distance was 1″·5, and in 1832 Sir John Herschel predicted that within the next year or two it would close up to such an extent that " none but the very finest telescopes will have any chance of showing this magnificent phenomenon." This prediction was verified in 1836, when the Dorpat refractor alone was able to elongate the star.

α Canum Venaticorum. 12h. 51m. + 38° 52′.

A very easy and interesting double, showing no signs at present of a binary character. Magnitude 3 and 6 ; distance 19″·8 ; position-angle 227°. This star was named Cor Caroli by Halley, at the suggestion of the court physician, who believed that it appeared more brilliant than usual on the evening before the return of King Charles II. to London.

ζ Ursæ Majoris. 13h. 20m. + 55° 27′.

Probably the best known double star in the sky, and certainly one of the easiest to find and most effective to look at in a small telescope. It is the middle star of the Bear's Tail. Magnitude 2·1 and 4·2 ; position-angle 147°·4 ; distance 14″·4 ; revolution very slow. The larger star of this pair is itself double, though telescopically single. It was the first discovered spectroscopic binary. Both components are bright, and of nearly equal magnitude. They revolve in 20 days 14 hours, in an eccentric orbit ; their combined mass is at least four times that of the Sun, and they are many times more luminous.

About 11′ away is the 5th magnitude star, Alcor, forming, with ζ Ursæ, the best-known example of a naked-eye pair.

α Virginis. 13h. 20m. − 10° 38′.

A first magnitude star whose spectrum is shifted backwards and forwards every four days by an amount denoting a revolution at the rate of 57 miles a second round the common centre of gravity of itself and a companion whose spectrum is just perceptible. Combined mass at least 2½ times Sun.

Globular Cluster ω Centauri. 13h. 21m. − 46° 57′.

The finest cluster of its kind in the sky, containing about 6,000 stars in a space of about 20′ diameter. Visible to the naked eye as a hazy comet-like object, giving as much light as a 4th magnitude star. The cluster contains 125 variable stars, of which 98 have periods less than 24 hours.

Cluster M. 3 Canum Venaticorum. 13h. 38m. + 28° 53′.

A very fine cluster of stars of the 11th magnitude and fainter, not very easily resolvable with small telescopes. This cluster is extraordinarily rich in variable stars ; no less than 132 out of 900 stars examined are regularly variable, many of them in very short periods.

α Boötis (Arcturus). 14h 11m. + 19° 42′.

The brightest star of the northern sky, with the very large proper motion of 2″·27 per year, which in the course of 1,600 years carries it across a space in the sky equal the apparent diameter of the Sun or Moon. Yet its parallax with respect to the fainter surrounding stars is small—only 0″·026 (Chase), whence it follows that its velocity at right angles to the line of light is about 200 miles a second, and its light is many hundred times that of the Sun.

α Centauri. 14h. 33m. – 60° 25′.

A splendid binary star. Components of the first magnitude ; distance 21‴·6 ; position-angle 211° (1902). Period 81 years. This is the nearest star to the Solar System, with a parallax 0″·75. The masses of the stars are very nearly equal, and one of them is in spectrum and in mass an almost precise counterpart of our Sun. The semi-major axis of the orbit is 23·6 times the length of the distance from Earth to Sun, or about a mean between the distances of Uranus and Neptune.

ε Boötis. 14h. 41m. + 27° 30′.

A most beautiful binary star. Magnitude 3, yellow, and 6½, blue ; distance 2‴·65 ; position-angle 328°·4 (1900.5).

ξ Boötis. 14h. 47m. + 19° 31′.

A very interesting binary star of great eccentricity of orbit. A, magnitude 4·5, yellow ; B, magnitude 6·5, purple. According to See's orbit, published in 1896, the period is 128 years ; but inasmuch as the place predicted from his orbit for 1903.5 was 154°·7, 1‴·25, whereas at that time it was measured as 186°·9, 2″·36, it is evident that this orbit requires modification.

Cluster M. 5 Libræ. 15h. 14m. + 2° 27′.

A globular cluster of faint stars remarkable for the number which are variable, about one in eleven, in periods mostly about 12 hours.

σ Coronæ. 16h. 11m. + 34° 7′.

An interesting binary star, with a period probably about 400 years. A, 6th magnitude, yellow ; B, 7th magnitude, bluish. Position-angle 210° ; distance 4″·38 (1900). The stars were at their closest about 1830, when their distance was only a little more than 1″ ; since that time they have been gradually opening out, and will continue to do so yet for about 100 years.

α Scorpii. 16h. 23m. – 26° 13′.

This fine reddish first magnitude star, the heart of the Scorpion, has a green 7th magnitude companion at a distance of 3″, in position-angle 270° ; but it is not easy to see on account of the glare of the bright star.

ζ Herculis. 16h. 38m. + 31° 47′.

A rapid binary star, which has performed more than three complete revolutions since it was discovered by Sir William Herschel on July 18th, 1782. A is 3rd magnitude, yellow ; and B is 6th magnitude, bluish. Its period is about 35 years, and the greatest separation of the components 1½". The companion passed periastron last in 1899 at a distance of 0"·5 ; since then it has opened out considerably, and a recent observation gives position-angle 205°·6, distance 1"·04 (1902·5). There is considerable evidence that the companion varies in colour from red to blue.

Globular Cluster M. 13 Herculis.   16h. 38m. + 36° 37'.

The finest globular cluster in the northern sky, and effective even in a small telescope, though the richest parts of the cluster can scarcely be resolved in the largest instruments. The whole contains at least 5,000 stars, of which only two of a thousand examined proved to be variable.

μ Scorpii.   16h. 45m. − 37° 53'.

Shown by the spectroscope to be a binary star with the short period of 34 hours 42 minutes. The two components have at maximum a relative velocity of nearly 300 miles a second ; this gives their separation as at least 6,000,000 miles ; and this their combined mass as at least 15 times that of the Sun.

α Herculis.   17h. 10m. + 14° 30'.

One of the finest coloured double stars. Magnitudes 2½, orange, and 6, blue ; distance 4"·88 ; position-angle 112°·1 (1901·5). The position-angle is very slowly diminishing.

Nebula M. 17 Sagittarii.   18h. 15m. − 16° 15'.

The Omega or Horse-shoe Nebula. This is one of the nebulæ that can be seen with comparatively small optical power. It is gaseous.

Nebula and Cluster M. 8 Sagittarii.   17h. 58m. − 24° 22'.

A magnificent irregular nebula in a very rich field of stars, too far south to be well seen in the latitude of England, where it rises only from 10° to 15° above the southern horizon. In a fine climate it is easily visible to the naked eye, and Gore speaks of it as " a glorious object with a 3-in. refractor in the Punjaub."

Planetary Nebula H. IV. 37 Draconis.   17h. 59m. + 66° 38'.

One of the most conspicuous planetary nebulæ in the sky, of a decided pale blue colour, looking, as do all such objects, very much like a star out of focus. Gaseous. This object lies close to the north pole of the ecliptic.

α Lyræ (Vega).   18h. 34m. + 38° 41'.

The second brightest star of the northern sky. A very white star, in whose atmosphere hydrogen is conspicuously absorbent. Proper motion 0"·36 ; annual parallax 0"·08 ; whence its light is about 100 times that of the Sun, and its velocity at right angles to the line of sight about 13 miles per second.

ε Lyræ.   18h. 41m. + 39° 30'.

A double star with components about 3' apart, separated to a good eye on clear moonless nights, and beautifully seen in an opera glass. A 3-in. telescope will show that each of the stars is itself a double, distance 2½″ and 3″. Between the two pairs are three smaller stars visible in a 4-in. telescope.

Nebula M. 57 Lyræ. 18h. 50m. + 32° 54′.

The famous Ring Nebula, very easy to find on the line between β and γ Lyræ. In a small telescope it appears as a faint ring, " a nebula with a hole in it " ; in large instruments it is seen that the central opening is not entirely clear of nebulosity. Exactly in the centre is a star which is very faint in the largest instruments, but which photographs comparatively easily. For the picture of it made with a very powerful photographic telescope see Plate 77.

β Cygni. 19h. 27m. + 27° 45′.

The finest coloured double star in the sky for small telescopes. Magnitudes 3, yellow, and 5½, blue. Distance 34‴·2 ; position-angle 55°·2.

α Aquilæ (Altair). 19h. 46m. + 8° 36′.

Magnitude 0·95 ; parallax 0‴·23 ; whence the light is about eight times that of the Sun.

α Capricorni. 20h. 13m. − 12° 51′.

A fine pair of stars of the 3rd and 4th magnitude, about 6′ apart, and easily separable to the naked eye. The preceding star has a 9th magnitude companion, distant about 45″ in position-angle 221° ; and the following a 9th magnitude companion, distant 154″ in position-angle 156°. There are also several faint closer companions to these stars, and the whole form a very fine group.

γ Delphini. 20h. 42m. + 15° 46′.

An easy double star. Magnitudes 4, yellowish, and 5, bluish ; distance 11‴·2 ; position-angle 270°·6 ; relative motion, if any, very slow.

61 Cygni. 21h. 2m. + 38° 13′.

A double star. Magnitudes 5·3 and 5·9 ; distance 22″ ; position-angle 125° (1900). Famous as the first star whose distance was determined. The pair has a very large common proper motion of over 5″ per year, which pointed it out as probably near to the Solar System. The mean of the best determinations of its parallax is 0‴·39 ; it is therefore, with one rather doubtful exception, the nearest star in the northern sky. Its light takes nearly 8½ years to reach us, and the two stars together give only about one-tenth the light of the Sun.

μ Cygni. 21h. 40m. + 28° 18′.

A fine double star, probably a binary, a good test for a small telescope. Magnitudes 4 and 5 ; distance 2‴·60 ; position-angle 122°·1 (Lewis, 1901·9). There is a companion, magnitude 7½ ; distant 209″, in position-angle 57°, which does not form part of the system. Since they were observed by Sir William Herschel in 1779, the pair has closed up from a separation of 7″ to its present distance.

ζ Aquarii.   22h. 24m. – 0° 32′.

A well-known and striking double star, easy to find in the centre of a triangle of naked-eye stars ; probably binary of long period.   Magnitudes 4 and 4 ; distance 3″·1 ; position-angle 321° (1899·9, Maw).

δ Cephei.   22h. 25m. + 57° 54′.

A remarkable double star, of which the brighter component is variable from 3·7m. to 4·9m. in a period of 5d. 8h. 48m., and is a spectroscopic binary of the same period.   But the variation of light is not due to an eclipse, since at the time of minimum the motion in the line of sight is at a maximum.   The variation is nevertheless undoubtedly due to the influence of the dark companion, possibly something of the nature of a tidal disturbance.   This star is typical of quite a numerous class of variable stars of short period, which are probably all spectroscopic binaries of the same type.   All have the characteristic that the rise in light is much quicker than the decline.

σ Cassiopeiæ.   23h. 54m. + 55° 12′.

A fine double star.   A, magnitude 5, white, and B, magnitude 7½, blue.   Position-angle 324° ; distance 3″·0.

PLATE 83.

STANDARD TIME.

As soon as communication by railway and telegraph is established in a country, it is convenient to adopt throughout the country a uniform system of time. Very usually the time adopted has been at first the mean time of the capital. But as communication between different countries increases, great inconvenience arises when allowance has to be made for a difference of adopted time involving an odd number of minutes and seconds. A large number of countries and states have therefore adopted a standard system of time based upon that of Greenwich, and differing from it by an exact number of hours, with occasionally an odd half hour.

Plate 83 shows the system of standard time adopted throughout the world, so far as it depends upon Greenwich. In Europe the time is generally that of Greenwich or one hour fast of it. Quite recently France has prepared to adopt Greeewich Time, and the only countries not included in the system are Portugal, Russia, Turkey, and Greece. The time one hour fast on Greenwich is known as Mid. Europe Time; that two hours fast as Eastern Europe Time.

In the United States and Canada there are five divisions. Inter-Colonial or Atlantic Time is four hours slow, Eastern or New York is five hours, Central six hours, Mountain seven hours, and Pacific Time eight hours slow on Greenwich.

On the 180th meridian the time is 12 hours different from Greenwich, and provision has to be made for the change of date. Since it would be very inconvenient to use a different date in different islands of the same group, the " date line" does not follow exactly the 180th meridian, but is drawn in a zig-zag course to avoid land. The greatest departure from the meridian is in the North Pacific Ocean, where the line takes a wide sweep west to give the Aleutian Islands the American date and then turns sharply eastward of the 180th meridian to avoid the extreme eastern portion of Siberia. To the east of the line the date prevails which has come round via America; to the west the date that has come by the Old World. Thus in the extreme east of Siberia the date is more than a day ahead of that in the Aleutian Islands.

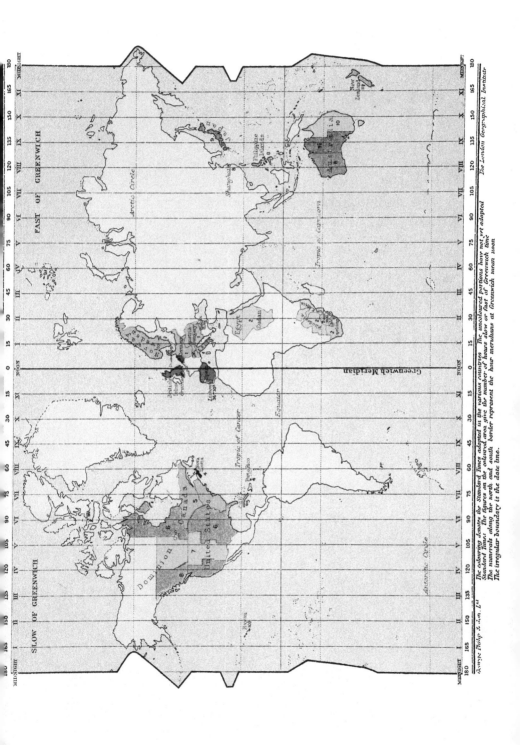

SLOW OF GREENWICH   EAST OF GREENWICH

Greenwich Meridian

The colouring denotes the Standard Times adopted in the various countries. The uncoloured portions have not yet adopted Standard Time. The figures on the coloured area give the number of hours slow or fast of Greenwich time. The numerals along the north and south border represent the hour meridians at Greenwich mean noon. The irregular boundary is the date line.

George Philip & Son, L.td

The London Geographical Institute

INDEX.

PAGE.

ABENEZRA, Lunar Object No. 88, Pl. 26 30

ABERRATION.—An apparent displacement of a star, arising from the progressive movement of light combined with the orbital movement of the Earth.

ABNEY, Sir W., Photograph of Solar Corona, Pl. 16...... 19

ABULFEDA, Lunar Object No. 85, Pl. 26, 31...... 30

ACHROMATIC.—Applied to a combination of lenses which conduct rays of different colours to the same focus.

ADAMS, Lunar Object No. 17, Pl. 26............... 30

ÆSTUUM SINUS, in Moon, N, Pl. 24, 32, 33 33

AGATHARCHIDES, Lunar Object No. 346, Pl. 25, 35 32

AGRIPPA, Lunar Object No. 244, Pl. 23, 31 31

AIRY, Lunar Object No. 106, Pl. 26, 31............ 30

ALBATEGNIUS, Lunar Object No. 109, Pl. 26, 31 30

ALCYONE, in Pleiades, Pl. 80......................62, 72

ALDEBARAN or α TAURI, Pl. 59 73

ALEXANDER, Lunar Object No. 209, Pl. 23, 27, 31 31

ALFRAGANUS, Lunar Object No. 77, Pl. 26......25, 30

ALGOL, or β PERSEI, remarkable Variable Star, Pl. 53 ....51, 72

ALGOL VARIABLES 51, 52

ALHAZEN, Lunar Object No. 151. Pl. 23, 27...... 31

ALIACENSIS, Lunar Object No. 97, Pl. 26, 31 ... 30

ALMANON, Lunar Object No. 86, Pl. 26, 31 30

ALPETRAGIUS, Lunar Object No. 267, Pl. 25, 32, 33 31

ALPHONSUS, Lunar Object No. 265, Pl. 25, 32, 33 31

ALPINE VALLEY, GREAT, Lunar Object No. 211, Pl. 23, 32 31

ALPS, Lunar Mountains, a, Pl. 23, 24, 32, 33, 34, 35...... 33

ALTAI, Lunar Mountains, h, Pl. 30 33

ALTITUDE.—The elevation of a body above the horizon, expressed in angular measure.

ANAXAGORAS, Lunar Object No. 418, Pl. 24, 33 32

ANAXIMANDER, Lunar Object No. 431, Pl. 24, 36 32

ANAXIMENES, Lunar Object No. 421, Pl. 24, 35, 36...... 32

PAGE.

ANDERSON, T. D., Discovery of Nova Persei ... 63

ANDROMEDA, Pl. 52, 71.

ANDROMEDA, Great Nebula in, M. 31, Pl. 18, 52, 74...... 21, 59, 70

ANDROMEDÆ, R, regularly Variable Star, Pl. 52 41

ANDROMEDÆ, Nova 1885, Pl. 52 54

ANDROMEDÆ, γ, a Double Star, Pl. 52, 53 ... 71

ANDROMEDIDS, Meteor Shower 5, 55

ANNUAL PARALLAX, Pl. 1 3

ANNULAR ECLIPSE OF SUN, Pl. 14 17

ANNULAR NEBULA, M. 57, in Lyra, Pl. 57, 78 61

ANOMALY.—The angle subtended at the Sun by a Planet, and the point of its orbit nearest the Sun, called the perihelion.

ANSGARIUS, Lunar Object No. 6, Pl. 26............ 30

ANTARES or α SCORPII, Pl. 68 78

ANTLIA, Pl. 66, 67.

APENNINES, Lunar Mountains, c, Pl. 20, 24, 32, 33, 34, 35 ......25, 33

APERTURE.—When applied to a telescope, means the diameter of the object glass.

APHELION.—The point of a Planet's orbit which is furthest from the Sun 5

APIANUS, Lunar Object No. 99, Pl. 26, 31 30

APOGEE.—The point of the Moon's orbit which is most distant from the Earth.

APOLLONIUS, Lunar Object No. 136, Pl. 23, 27, 28 30

APSE.—In a planetary orbit the apses are the points otherwise known as perihelion and aphelion 4

APUS, Pl. 68, 70.

AQUARII, ζ, a Double Star, Pl. 63 81

AQUARIUS, Pl. 63, 64, 69, 71, 72................... 3

AQUILA, Pl. 62, 63, 71, 72.

AQUILÆ, α, (Altair) 80

AQUILÆ, η, regularly Variable Star, Pl. 62 53

AQUILÆ, Nova 1899 54

ARA, Pl. 68, 72.

ARAGO, Lunar Object No. 250, Pl. 23, 30......... 31

ARATUS, Lunar Object No. 238, Pl. 23 31

ARCHIMEDES, Lunar Object No. 399, Pl. 24, 32, 33, 34, 35...... 25, 32

Printed in the United States
By Bookmasters